Praise for *The B*

'*The Burning Question* is a fascinating examination of the forces that have led to our current predicament and it presents an important framework for a sustainable future. I recommend it highly. The climate crisis is a challenge unprecedented in its scale and complexity. We simply must confront this existential challenge and stop making it worse. That will require the awakening and activism of people all around the world.'

Al Gore
45th Vice-President of the United States

'The issues explored in *The Burning Question* are hugely important. Policymakers and the public urgently need to be engaging in this kind of big-picture conversation.'

Jim Hansen
Director of the NASA Goddard Institute for Space Studies

'This is a book that needed to be written: it asks the right question then seeks the most effective ways of answering it. An essential contribution to our thinking about climate change.'

George Monbiot
Writer and campaigner

'Fossil fuels are so last century. *The Burning Question* tells us clearly why and how to get off them, but crucially also explores why we aren't doing anything much about it at the moment, and points the finger at the villains of the piece. Terrific.'

Sir Tim Smit
Founder of the Eden Project

'*The Burning Question* is one of those books that doesn't shy away from delivering an uncomfortable message – there's no sweetening of the pill, placating political interests or pandering to commercial sensibilities – it simply tells it like it is. But much more than that, in accessible language it develops responses to the challenges we face – not utopian social change, or unrealistic technical wizardry, but rather a portfolio of options thought through at a system level. *The Burning Question* is an important contribution to understanding both the scale of the climate challenge and how we may yet develop a low-carbon and climate-resilient society.'

Professor Kevin Anderson

Deputy director, Tyndall Centre for Climate Change research

'To keep climate denial from turning into climate despair – that we don't know how to solve the climate challenge without suppressing civilisation – we need a realistic assessment of the problem and an optimistic set of solutions. This book gives us both, in a short but compelling narrative that may be the difference between a glide to a decent future and a crash of civilisation. Read it, share it, and start preaching its gospel.'

Durwood Zaelke

President of the Institute for Governance and Sustainable Development

'An extremely clear-sighted and highly readable account of the factors fanning the flames of climate change with plenty of practical suggestions how to set about extinguishing them.'

Baroness Worthington

Climate change policy expert and life peer

'It's terrifyingly simple. Burning carbon made our modern industrial world. Now we've got to stop burning it. We've got to stop drilling for oil and gas, and leave the coal in the ground. We've got to prick the carbon bubble, write off half the assets of the world's biggest industry, and break the infrastructure and mental lock-in that is preventing viable new energy technologies from taking over. This is the big-picture story of why and how that must happen. And why, so far, we are abjectly failing. Brilliant.'

Fred Pearce
Author of *The Last Generation*

'At a time when we're making the climate debate "small", a series of bite-sized chunks each to be "smuggled" through a resistant policy system, Berners-Lee and Clark remind us that the debate is actually huge in its global scope, its likely impact and, most importantly of all, in terms of the solutions we need to adopt.'

Mike Barry
Head of Sustainable Business, Marks & Spencer

'An easy-to-read book about a difficult-to-solve problem. Berners-Lee and Clark illustrate why climate change is such a complex issue. But also that it has a solution.'

Samuel Fankhauser
Co-Director, Grantham Research Institute on Climate Change and the Environment, LSE

'The image of scientists and academics used to be one of calm, mild-mannered people but today the frustration among many is palpable. This book shows why. The gap between evidence, policy and practice is yawningly wide.

This book tries to bridge that gap, offering a reasoned account of the problem and suggesting what we might do about it – from global policy to culture change.'

Tim Lang
Professor of food policy, City University London

'Climate change is the most difficult problem the world has ever faced. Berners-Lee and Clark have compressed this complex issue into a short and highly readable book that covers science, psychology and sociology. Uncompromisingly rigorous but easy to read, this book is a perfect introduction to the central topic of the twenty-first century.'

Chris Goodall
Low-carbon technology expert and author of *Sustainability: All That Matters*

'This book hits the climate nail bang on the head: we can only avoid devastating damage if most of the world's coal, oil and gas are left in the ground. In wonderfully clear and readable prose, the authors set out the facts and what we must do about them. It deserves to be widely read: I only hope it will reawaken the climate movement, which has gone into such desperate decline over the last three years. Only public pressure will force governments to close down coal fired power stations and end our oil dependence: this book is a lucid and powerful call to arms.'

Michael Jacobs
Visiting professor, Grantham Research Institute, LSE and former special adviser on climate change to the UK Prime Minister

THE BURNING QUESTION

MIKE BERNERS-LEE
DUNCAN CLARK

THE BURNING QUESTION

We can't burn half the world's oil, coal and gas. So how do we quit?

David Suzuki Foundation
GREYSTONE BOOKS
Vancouver/Berkeley

For our kids and theirs

First published in Great Britain in 2013 by Profile Books Ltd.,
3a Exmouth House, Pine Street, London EC1R 0JH
www.profilebooks.com

13 14 15 16 17 5 4 3 2 1

Greystone Books Ltd.
www.greystonebooks.com

David Suzuki Foundation
219–2211 West 4th Avenue
Vancouver BC Canada V6K 4S2

Cataloguing data available from Library and Archives Canada
ISBN 978-1-77164-007-7 (pbk.)
ISBN 978-1-77164-008-4 (epub)

Front cover design by Peter Dyer
Text design by Duncan Clark and Henry Iles
Cover illustration: iStockphoto.com
Printed and bound in Canada by Friesens
Distributed in the U.S. by Publishers Group West

We gratefully acknowledge the financial support of the Canada Council for the Arts, the British Columbia Arts Council, the Province of British Columbia through the Book Publishing Tax Credit, and the Government of Canada through the Canada Book Fund for our publishing activities.

Greystone Books is committed to reducing the consumption of old-growth forests in the books it publishes. This book is one step towards that goal.

Contents

Part four: Not just fossil fuels

The other ways we're warming the planet, such as soot from cooking fires and methane from livestock. How efforts to reduce these other drivers of climate change will be crucial to what happens in the next few decades.

Part five: What now?

Six key steps that will help tackle climate change.

Do the maths

Foreword by Bill McKibben

If the pictures of towering wildfires, devastating droughts and crippling hurricanes haven't convinced you, here are some hard numbers about climate change. May 2012 was the hottest month on record for the Northern Hemisphere – the 327th consecutive month in which the temperature of the entire globe exceeded the twentieth-century average, the odds of which occurring by simple chance were 3.7×10^{99}, a number considerably larger than the number of stars in the universe.

The June that followed broke or tied 3,215 high-temperature records across the United States, hot on the heels of America's warmest ever spring, which crushed the old record by so much that it represented the 'largest temperature departure from average of any season on record'. The same week, Saudi authorities reported that it had rained in Mecca despite a temperature of 109 degrees, the hottest downpour in the planet's history. In the autumn, a hurricane of unprecedented power slammed into the New York City region, causing tens of billions of dollars in damage. As the year ended, England announced it had suffered its wettest year ever recorded and Australia entered a hot spell that became so severe its weather service had to add two extra colours to its temperature maps.

Not that our leaders seem to notice. The meeting in Rio for the twentieth-anniversary reprise of a massive 1992 environmental summit accomplished nothing. Unlike George H. W. Bush, who flew in for the first conclave, Barack Obama didn't even attend. It was 'a ghost of the glad, confident meeting twenty years ago,' journalist George Monbiot wrote; no one paid it much attention, footsteps echoing through the halls 'once thronged by multitudes.' Since I wrote one of the first books for a general audience about global warming way back in 1989, and since I've spent the intervening decades working ineffectively to slow that warming, I can say with some confidence that we're losing the fight, badly and quickly – losing it because, most of all, we remain in denial about the peril that human civilisation is in.

When we think about global warming at all, the arguments tend to be ideological, theological and economic. But to grasp the seriousness of our predicament, you just need to do a little maths. Recently, an easy and powerful bit of arithmetical analysis first published by financial analysts in the UK has been making the rounds of environmental conferences and journals, but it hasn't yet broken through to the larger public. This analysis upends most of the conventional political thinking about climate change. And it allows us to understand our precarious – our almost-but-not-quite-finally hopeless – position with three simple numbers.

The first number: 2° Celsius

If the movie had ended in Hollywood fashion, the Copenhagen climate conference in 2009 would have marked the culmination of the global fight to slow changing climate. The world's nations had gathered in the December gloom of the Danish capital for what a leading climate economist, Sir Nicholas Stern, called the 'most important gathering since the Second World War, given what is at stake.' As Danish energy minister Connie Hedegaard, who presided over the conference, declared at the time: 'This is our chance. If we miss it, it could take years before we get a new and better one. If ever.'

In the event, of course, we missed it. Copenhagen failed spectacularly. Neither China nor the United States, which between them are responsible for 40 per cent of global carbon emissions, was prepared to offer dramatic concessions, and so the conference drifted aimlessly for two weeks until world leaders jetted in for the final day. Amid considerable chaos, President Obama took the lead in drafting a face-saving 'Copenhagen Accord' that fooled very few. Its purely voluntary agreements committed no one to anything, and even if countries signalled their intentions to cut carbon emissions, there was no enforcement mechanism.

The accord did contain one important number, however. In Paragraph 1, it formally recognised 'the scientific view that the increase in global temperature should be below two degrees Celsius'. And in the very next paragraph, it declared that 'we agree that deep cuts in global emissions are required ... so as to hold the increase in global temperature below two degrees Celsius.' By insisting on two degrees – about 3.6 degrees Fahrenheit – the accord ratified positions taken earlier in 2009 by the G8, and the so-called Major Economies Forum. It was as conventional as conventional wisdom gets. The number first gained prominence, in fact, at a 1995 climate conference chaired by Angela Merkel, then the German minister of the environment and now the centre-right chancellor of the nation.

Some context: so far, we've raised the average temperature of the planet just under 0.8 degrees Celsius, and that has caused far more damage than most scientists expected. (A third of summer sea ice in the Arctic is gone, the oceans are thirty per cent more acidic, and since warm air holds more water vapour than cold, the atmosphere over the oceans is a shocking five per cent wetter, loading the dice for devastating floods.) Given those impacts, in fact, many scientists have come to think that two degrees is far too lenient a target. 'Any number much above one degree involves a gamble,' writes Kerry Emanuel of MIT, a leading authority on hurricanes, 'and the odds become less and less favourable as the temperature goes up.' Thomas Lovejoy, once the World Bank's chief biodiversity adviser, puts it like this:

'If we're seeing what we're seeing today at 0.8 degrees Celsius, two degrees is simply too much.'

Despite such well-founded misgivings, political realism bested scientific data, and the world settled on the two-degree target – indeed, it's fair to say that it's the only thing about climate change the world has settled on. All told, 167 countries responsible for more than 87 per cent of the world's carbon emissions have signed on to the Copenhagen Accord, endorsing the two-degree target. Only a few dozen countries have rejected it, including Kuwait, Nicaragua and Venezuela. Even the United Arab Emirates, which makes most of its money exporting oil and gas, signed on. The official position of planet Earth at the moment is that we can't raise the temperature more than two degrees Celsius – it's become the bottomest of bottom lines. Two degrees.

The second number: 565 gigatonnes

Scientists estimate that humans can pour roughly 565 more gigatonnes of carbon dioxide into the atmosphere by midcentury and still have some reasonable hope of staying below two degrees. ('Reasonable,' in this case, means four chances in five, or somewhat worse odds than playing Russian roulette with a six-shooter.) This number isn't exact, but few dispute that it's generally right. It was derived from one of the most sophisticated computer-simulation models that have been built by climate scientists around the world over the past few decades. And the number is being further confirmed by the latest climate-simulation models in advance of the next report by the Intergovernmental Panel on Climate Change (IPCC). 'Looking at them as they come in, they hardly differ at all,' says Tom Wigley, an Australian climatologist at the National Center for Atmospheric Research. 'There's maybe forty models in the data set now, compared with twenty before. But so far the numbers are pretty much the same. We're just fine-tuning things. I don't think much has changed over the last decade.' William Collins, a senior climate scientist at the Lawrence Berkeley National Laboratory, agrees. 'I think the results of this round of simulations will be quite similar,' he says. 'We're not getting any free lunch from additional understanding of the climate system.'

We're not getting any free lunch from the world's economies, either. With only a single year's lull in 2009 at the height of the financial crisis, we've continued to pour record amounts of carbon into the atmosphere, year after year. The International Energy Agency's (IEA) latest figures showed that carbon dioxide emissions rose to 31.6 gigatonnes in 2011, up 3.2 per cent from the year before. America had a warm winter and converted more coal-fired power plants to natural gas, so its emissions fell slightly; China kept booming, so its carbon output (which recently surpassed the US) rose 9.3 per cent; the Japanese shut down their fleet of nukes post-Fukushima, so their emissions edged up 2.4 per cent. 'There have been efforts to use more renewable energy and improve energy efficiency,' said Corinne Le Quéré, who runs England's Tyndall Centre for Climate Change Research. 'But what this shows is that so far the effects have been marginal.' In fact, study after study predicts that carbon emissions will keep growing by roughly 3 per cent a year – and at that rate, we'll blow through our 565-gigatonne allowance in sixteen years, around the time today's preschoolers will be graduating from high school. 'The new data provide further evidence that the door to a two-degree trajectory is about to close,' said Fatih Birol, the IEA's chief economist. In fact, he continued, 'When I look at this data, the trend is perfectly in line with a temperature increase of about six degrees.' That's almost 11 degrees Fahrenheit, which would create a planet straight out of science fiction.

So, new data in hand, everyone at the Rio conference renewed their ritual calls for serious international action to move us back to a two-degree trajectory. The charade continued in November when the latest Conference of the Parties (COP) of the UN Framework Convention on Climate Change convened in Qatar. That was COP 18. COP 1 was held in Berlin in 1995, and since then the process has accomplished essentially nothing. Even scientists, who are notoriously reluctant to speak out, are slowly overcoming their natural preference to simply provide data. 'The message has been consistent for close to thirty years now,' Collins says with a wry laugh, 'and we have the instrumentation and the computer power required to present the evidence in detail. If we choose to continue on our present course of action, it should be done with a full evaluation of the evidence the scientific community has presented.' He pauses, suddenly conscious of being on the record. 'I should say, a fuller evaluation of the evidence.'

So far, though, such calls have had little effect. We're in the same position we've been in for a quarter-century: scientific warning followed by political inaction. Among scientists speaking off the record, disgusted candour is the rule. One senior scientist told me, 'You know those new cigarette packs, where governments make them put a picture of someone with a hole in their throats? Gas pumps should have something like that.'

The third number: 2,795 gigatonnes

This number is the scariest of all – one that, for the first time, meshes the political and scientific dimensions of our dilemma. It was brought to wide attention first by the Carbon Tracker Initiative, a team of London financial analysts and environmentalists who published a report in an effort to educate investors about the possible risks that climate change poses to their stock portfolios. The number describes the amount of carbon already contained in the proven coal and oil and gas reserves of the fossil-fuel companies, and the countries (think Venezuela or Kuwait) that act like fossil-fuel companies. In short, it's the fossil fuel we're currently planning to burn. And the key point is that this new number – 2,795 – is higher than 565. Five times higher.

The Carbon Tracker Initiative combed through proprietary databases to figure out how much oil, gas and coal the world's major energy companies hold in reserve. The numbers aren't perfect – they don't fully reflect the recent surge in unconventional energy sources like shale gas, and they don't accurately reflect coal reserves, which are subject to less stringent reporting requirements than oil and gas. But for the biggest companies, the figures are quite exact: If you burned everything in the inventories of Russia's Lukoil and America's ExxonMobil, for instance, which lead the list of oil and gas companies, each would release more than 40 gigatonnes of carbon dioxide into the atmosphere.

Which is exactly why this new number, 2,795 gigatonnes, is such a big deal. Think of two degrees Celsius as the legal drinking limit – equivalent to the 0.08 blood-alcohol level below which you might get away with driving home. The 565 gigatonnes is how many drinks you could have and still stay below that limit – the six beers, say, you might consume in an evening. And the 2,795 gigatonnes? That's the three 12-packs the fossil-fuel industry has on the table, already opened and ready to pour.

We have five times as much oil and coal and gas on the books as climate scientists think is safe to burn. We'd have to keep 80 per cent of those reserves locked away underground to avoid that fate. Before we knew those numbers, our fate had been likely. Now, barring some massive intervention, it seems certain.

Yes, this coal and gas and oil is still technically in the soil. But it's already economically aboveground – it's figured into share prices, companies are borrowing money against it, nations are basing their budgets on the presumed returns from their patrimony. It explains why the big fossil-fuel companies have fought so hard to prevent the regulation of carbon dioxide – those reserves are their primary asset, the holding that gives their companies their value. It's why they've worked so hard these past years to figure out how to unlock the oil in Canada's tar sands, or how to drill miles beneath the sea, or how to frack the Appalachians.

If you told Exxon or Lukoil that, in order to avoid wrecking the climate, they couldn't pump out their reserves, the value of their companies would plummet. John Fullerton, a former managing director at JP Morgan who now runs the Capital Institute, calculates that at today's market value, those 2,795 gigatonnes of carbon emissions are worth about $27 trillion. Which is to say, if you paid attention to the scientists and kept 80 per cent of it underground, you'd be writing off $20 trillion in assets. The numbers aren't exact, of course, but that carbon bubble makes the housing bubble look small by comparison. It won't necessarily burst – we might well burn all that carbon, in which case investors will do fine. But if we do, the planet will crater. You can have a healthy fossil-fuel balance sheet, or a relatively healthy planet – but now that we know the numbers, it looks like you can't have both. Do the maths: 2,795 is five times 565. That's how the story ends.

A longer version of this piece originally appeared in Rolling Stone *magazine.*[1]

Introduction

The aim of this book is to give a very big perspective on a very big challenge. Much of it revolves around two key messages that are widely ignored. The first of these is that avoiding unacceptable risks of catastrophic climate change means burning less than half of the oil, coal and gas in currently commercial reserves – and a much smaller fraction of all the fossil fuels under the ground. This isn't a new realization. NASA scientist James Hansen and colleagues first pointed it out many years ago and as we were writing this book environmentalist Bill McKibben blasted the idea to a wider audience in the article excerpted above, the original of which became an unexpected viral hit for the *Rolling Stone* website. Our book picks up where McKibben signs off, exploring the implications of the world's fossil fuel abundance. Who owns it all? What's it worth? Is it really true that we can only burn a fifth, or that stock exchanges are exposed to a 'carbon bubble'? Could we not burn the fuel and capture the carbon? How does all this affect the geopolitics of the issue?

This book's second big message is that viewed at the global level – the *system* level – many of the things we assume will help reduce fossil fuel use turn out to make much less of a difference than common sense would suggest. While clean energy, greener behaviour, energy efficiency and slowing population growth all have a key role to play, they don't appear to reduce the ever-increasing rate at which fossil fuels come out of the ground. A close look at the statistics shows that man-made carbon emissions have followed a long-term exponential trend that has proved uncannily resilient to social or technical change. There are powerful feedback mechanisms at work in human energy use capable of cancelling out apparent green progress. It's like squeezing a balloon: gains made in one place reappear as bulges elsewhere.

When you put these two messages together, and look at them in the light of the latest climate science, the inescapable finding is that we need to deal with the fossil fuel problem head on. In other words, there's no safe alternative to deliberately and rapidly constraining the rate at which fossil fuels come out of the ground and flow through the global economy. One day technology and infrastructure may exist to allow us to burn the world's oil, coal and gas safely by capturing and burying the carbon emissions. That technology needs much more effort put into it, but on current trends remains a long way off. In the meantime, the choice we face is between taking unimaginable risks with the planet and leaving vastly valuable fossil fuels in the ground.

Given that stark choice, and our inability to dent the emissions trajectory so far, it seems critical that we get a proper understanding of the core barriers that are holding us back. In the chapters following, we explore the economic, psychological, cultural and political roots of our inertia. We also show how the money bound up in oil, coal and gas reserves has blocked political progress and clouded the analysis. The fossil fuel sector is often described as the biggest and most profitable industry of all time, and it's currently working hard not just to resist carbon cuts but to grow its reserves and markets as fast as possible. The world needs to send it – and its investors – a loud and clear message: stop!

But it's not good enough to pin all the blame on fossil fuel companies as there are many other barriers to progress. We look at the rate at which the world is sinking money into vehicles, power plants and other infrastructure that require fossil fuels to run; the nervousness of policymakers to do anything that will push up energy prices; and perhaps most importantly of all, the host of social and psychological traits that have been stopping our species from waking up to the threat we face.

While fossil fuel is the key driver of climate change, humans also tamper with the atmosphere by releasing a host of other greenhouse gases and particles, many of which come from agriculture and deforestation. We look at how rapid action here can increase the chance of successfully tackling climate change and buy us some time to get fossil fuels under control. The way the world's land is managed over the coming decades will be crucial not just for feeding the growing population, but for cutting emissions, capturing carbon, protecting biodiversity and even, if humanity plays its cards right, responsibly liberating some land for producing energy.

Finally, we outline what we think needs to be done. We argue that tackling the problem will require a mix of harder global politics, passionate campaigning and smarter use of technology – each of which needs to support the others. We address the critical challenge of collectively waking up from decades of slumber. And we look at leadership, too often absent thus far, describing how everyone can contribute.

Some of the facts about climate change are uncomfortable but this book is about clarity so we have deliberately avoided playing down the severity of the situation just to keep the message palatable for our readers. But nor have we feigned optimism, so when you do see glints of positivity, they too are real.

A note about notes

Climate change is surely the most multi-layered and intriguing problem in history – a perfect storm of money and power, science and politics, technology and the human mind. In order to address and join the dots between these many elements, we have focused unapologetically on the big picture: the core perspectives that too often get lost amid detailed discussion. If you want chapter and verse on the economics, politics or psychology of climate change, or on alternative energy sources, then you'll need to look elsewhere. A good starting point would be our endnotes, which are designed to be read and often include further discussion as well as links to sources and suggested reading. In addition to appearing at the back of the book, they are posted online, along with some of the key datasets that we've used, at burningquestion.info – where readers can also get in touch with us to share reactions and ideas.

PART ONE

A problem of abundance

Why burning all the world's oil, coal and gas is not an option

1. The curve

Fossil fuel use and emissions have been rising exponentially for more than a century

Nothing has shaped human history more than our use of energy. It was energy from the sun that provided the conditions for life on earth. It was the competition for food energy that determined the winners and losers of evolution. It was energy from fire that enabled early humans to cook, radically improving their diets and enabling their energy-hungry brains to grow even bigger.[1] Bigger brains eventually led to agriculture – a systematic means of harnessing the sun's energy – which in turn freed us from the constraints of nomadic existence and gave rise to permanent settlements.

In villages and towns, the human population flourished – as did knowledge. More people with more ideas unlocked yet more sources of energy. Domesticated animals pulled ploughs and dug irrigation channels, enabling us to exploit more land. Forests – essentially banks of stored solar energy – were cleared first to produce firewood and building materials, then to make charcoal, a higher-grade form of energy that burned hot enough to smelt the bronze for the bronze age and the iron for the iron age. Better materials led to better tools and technology, which soon unlocked yet more types of energy. Mills started harnessing the kinetic energy of rivers to grind grain, operate mechanical hammers and even make cloth. Wind energy was harvested too, not just by mills but also by sailing ships, opening up long-distance trading, both in goods and knowledge – including knowledge about energy.

More energy, more technology, more people, more energy – this age-old feedback loop was the motor that enabled humankind to dominate the earth. From the earliest days, it brought environmental side effects. As our numbers ballooned we hunted many species to extinction. As we dammed rivers for watermills we killed off migratory fish. But for most of human history, the earth coped reasonably well with the onslaught its brainiest and most energy-hungry inhabitant was wreaking. That started to change in the eighteenth century when people accessed significant quantities of fossil fuels. While burning wood had allowed us to release solar energy captured over years, decades or centuries, fossil fuels contained hundreds of millions of years' worth of sunlight. The energy was stored in carbon-based molecules formed when ancient plants decomposed anaerobically under layers of sediment to form incredibly energy-dense solids, liquids and gases: coal, crude oil and natural gas.

Small-scale fossil fuel use began much earlier. Coal was burned in ancient Greece, Roman Britain, Aztec Mexico and imperial China – though it remained a novelty to Marco Polo who wrote with amazement during a thirteenth-century trip to China about 'a kind of black stones existing in beds in the mountains, which ... burn like firewood.'[2] The reason fossil fuels remained so obscure for so long was that most of them were trapped deep underground. Finding them was difficult and getting them out was even harder.

The steam engine solved that problem. It used the incredible power of coal to drain water from deep mines, making it possible to access more coal to power more steam engines. First in Britain, then in Europe, Japan and elsewhere, the energy–society feedback loop went into overdrive. Coal enabled railways, steam ships, blast furnaces, brick kilns and metal smelters to multiply. It increased the supply of every commodity and material and boosted progress on every front of technology, from medicine to microscopy. As political economist William Stanley Jevons noted in 1865, energy from coal had become 'the universal aid ... the factor in everything we do.'[3] And that was before coal had enabled electricity to develop from a scientific curiosity into a major new industry, offering countless new ways to consume energy.

Importantly, coal didn't replace existing energy sources; it augmented them. Global consumption of energy from trees and other forms of 'biomass' hardly changed as industrial-scale coal mining spread around the world. In turn, use of coal kept growing as its liquid cousin – crude oil – rose to significance in the early twentieth century. With oil came motorised transport, which amplified energy use yet further. It allowed us to move around faster, trade with people further afield, and to spread into suburbs. Homes grew in both numbers and size, increasing demand for building materials, furniture, heating and electricity – all of which took yet more energy to provide. Food production was also transformed by oil, which powered tractors, trucks and factory fishing fleets as well as easing pressure on farmland by providing synthetic alternatives to cotton, wool and wood. With less need for manual labourers, the workforce shifted towards factories, many of which not only consumed energy directly but churned out machines such as cars, light bulbs and radios that themselves took energy to run.

Global use of oil and coal kept rising as the third major fossil fuel – natural gas – started to scale up after the Second World War. Gas boosted energy supply directly, ramping up electricity production and fuelling boilers and cookers. But it also drove up demand for every other kind of energy by enabling the continued expansion of the global population. As human numbers rose towards three billion in the 1950s, a catastrophic crunch in food production was avoided in large part thanks to huge quantities of nitrogen fertilisers produced from natural gas. As the population shot up, so did energy demand.

More energy, more technology, more people, more energy. The feedback loop kept whirring throughout the twentieth century as holiday flights, cars, washing machines and central heating became the norm in wealthy countries. Energy supply expanded too as large-scale hydroelectric and nuclear power plants started coming on stream in significant numbers. But human demand for energy showed no signs of hitting a ceiling so fossil fuel use kept rising too. Society produced more goods of every type and found new ways to use energy – from domestic air conditioning and space missions to kidney dialysis machines and, of course, computers, which turned out to be extremely helpful for finding yet more fossil fuel reserves and developing the tools for getting them out of the ground.

The effects of escalating energy use on the planet have been extraordinary. Today we total more than seven billion people and consume more than five-hundred billion billion joules of energy each year. That's comparable to each man, woman and child having more than a hundred full-time servants doing manual work on their behalf.[4] Aided by this vast army of energy slaves, we've transformed the planet so fundamentally that scientists are currently considering whether formally to call time on the Holocene geological era and recognise that we're now in the Anthropocene: the human era. A remarkable 40 per cent of the world's land surface is now farmed or grazed. Most of the large rivers are dammed to provide water for irrigation, washing and drinking. Seen from space, even the shadowy side of the planet glows visibly thanks to tens of billions of electric lights.

But perhaps the greatest adjustment that human energy use has made to the Earth remains invisible. Our fuels and farming have been reformulating the atmosphere by flooding the air with carbon that was once locked up in forests, coal, oil and gas. Nineteenth-century physicists theorised that all this extra carbon – in the form of carbon dioxide gas – would warm the planet by trapping heat that would otherwise escape to space. The science firmed up over the decades, though no one knew if the impacts would be positive or negative and few were sufficiently concerned to consider abandoning the fossil fuels that had taken humankind to such heights. So the energy–society feedback kept spinning and emissions of carbon dioxide kept climbing.

But to say that carbon emissions have been climbing for centuries doesn't tell the whole story. They have in fact been climbing in a very specific way: *exponentially.*

The carbon curve

An exponential curve has special properties. At any point, the steepness is proportional to the height. Not only that but the *rate* of increase is proportional to the steepness.[5] In other words, it's the type of curve you get when the more of something you have, the faster that something grows. Exponentials often turn up when there's a positive feedback loop at work. The population of insects in a jar follows an exponential curve for as long as there is adequate food supply because the more insects there are the faster they can breed. A credit card debt grows exponentially as the interest gets applied to ever more interest.

Exponential curves can't go on forever. They get steeper and steeper until eventually, something has to give. Sometimes they crash, like the population of the insects in the jar when the food runs out, or the credit card debt when bankruptcy is declared. Or sometimes they tame themselves more gently – such as the way global population has slowed to a steady rise over the past few decades (more on that later). But they can't go on for all time.

Although there is a bit of snaking around along the way, the long-term fit of global carbon emissions to an exponential curve is uncanny – as the graph overleaf shows. The most noticeable deviation is the slow-down in emissions growth in the first half of the twentieth century. It's an open question whether this was the result of two world wars, or the Great Depression, or global energy supplies struggling to keep up with demand. But

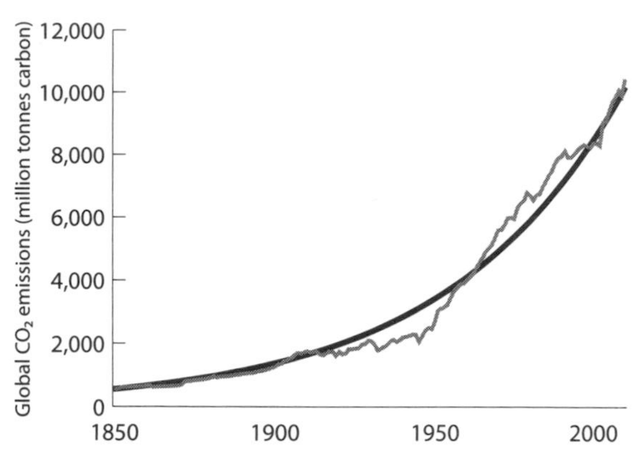

Annual man-made emissions of carbon dioxide from all major sources – fossil fuel use, cement production, deforestation and land use changes – plotted against an exponential line. The neatness of the correlation was first explored in a study by Andrew Jarvis and colleagues at Lancaster University.[6]

the carbon curve quickly got back on track in the 1950s, helped along by many huge oilfields coming on stream and a major acceleration in global population. Emissions then surged above the long-term trend in the energy-profligate 1960s before slowing in the wake of oil price spikes in the 1970s and early 1990s and finally getting back on track around the turn of the millennium.

It's tempting to assume that the last decade must surely have seen a bit of slowdown on the global carbon curve given all the green summits, hybrid cars and the low-energy light bulbs. Focusing on just a few years isn't statistically very meaningful, but for what it's worth the last decade tells the opposite story, lying *above* rather than below the long-term trend line. From 2000 to 2010, the average annual growth of carbon emissions from all man-made sources was around 2.3 per cent – higher than the very long-term trend of 1.8 per cent. The figures would look even worse was it not for savings in deforestation; fossil fuel emissions rose by a massive 3 per cent a year for the last decade

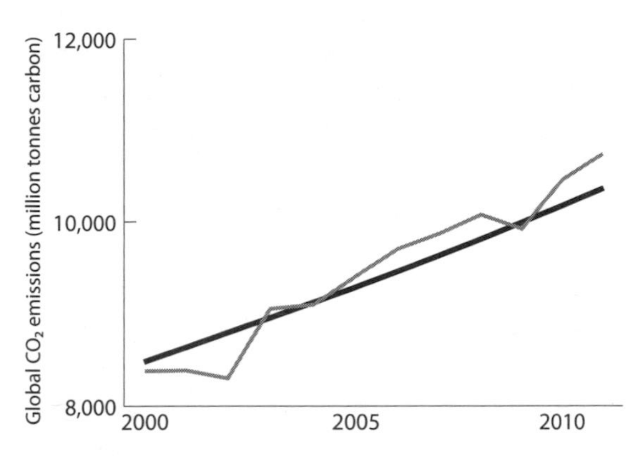

A close up of the same graph since 2000. In this period average emissions growth has been around 2.3 per cent – higher than the very long-term trend of 1.8 per cent. (Includes some estimates for deforestation and cement.)

and continued to do so in 2011, the last year for which reliable data is available at the time of writing.[7] As we go to press, it looks like 2012 saw a similar rise.

The unremittingly exponential nature of the global carbon curve fits perfectly with the idea that society's use of energy is driven by a powerful positive feedback mechanism. It also makes it painfully clear that none of the action taken so far to deal with climate change has succeeded in making the smallest difference at the global level. It is theoretically possible that without all the efficiency gains and corporate responsibility reports, the community action groups, global summits and so on, emissions might have been even higher: we might have deviated significantly above the long-term trend. That would require a coincidence but is not impossible. It is also conceivable that all those efforts have been successfully laying the groundwork for reduced global emissions in the future. But the striking reality is that any progress so far is invisible on the graph.

Interestingly, the simple and predictable global emissions trend belies a shifting picture in the sources of those emissions. The destruction of forests which dominated the human carbon footprint 160 years ago has been increasingly dwarfed first by coal and then by a combination of coal, oil and gas. And, in a rare good news story, emissions from deforestation have fallen sharply in absolute terms during the last two decades. But the curve has carried on as if nothing had changed. Similarly, at first glance the energy feedback loop appears to have finally burned out in rich countries, where energy use and emissions (and in some cases populations) are now falling. But again the global exponential trend has been unaffected. It looks as if the carbon curve exhibits a capacity for self-regulation, similar to the way a human body maintains a steady temperature almost regardless of what goes on in its surrounding environment. We'll explore this more in part two.

The unflinching resilience of the curve, decade after decade, bouncing back from wars and depressions, and despite endless technological revolutions, tells us a lot about the challenge we are up against. Not only are we faced with trying to slow down the proverbial oil tanker; we need to do so while it's accelerating harder than ever before, driven by what appears to be a deep dynamic in which access to energy begets yet more energy.

But as we've seen, exponentials can't go on for all time, and our carbon curve is no exception, as a few back-of-the-envelope calculations reveal. If we stayed on our current trajectory for, say, 600 years, emissions would rise so steeply that our fuel-burning would consume all the oxygen that is currently in the Earth's atmosphere every single year, leaving nothing at all for us to breath.[8] So we can safely say that we'll be coming off the curve well before 2600 AD, whether we like it or not. In reality, supplies of fossil fuels would run out much sooner than that. Chapter three takes stock of all the oil, coal and gas resources left in the world. The total suggests that if the current carbon curve continued, we'd most likely run out of all of them – including all the obscure and unconventional fuel sources such as tar sands and shale gas – within a century or so.

In other words, the carbon curve doesn't necessarily have *that* much life left in it. Although it's been running for hundreds of years, it most likely couldn't run for hundreds more – even if we didn't care about climate change. So our first really big question is, can we just let the curve run its 'natural' course and either crash or level off of its own accord as fossil fuels get more scarce and expensive and other energy sources take over? Or do we need to intervene to bring it to heel more quickly? The answer depends mainly on two other questions. How quickly will we run into trouble if we let things run their course? And how does that fit in with the amount of fossil fuels we have left?

2. Heading for trouble fast

Why the world's 2°C climate change target is both too risky and too tough

Before exploring why the curve keeps outsmarting us – and what it will take to tame it – we'll pause to check whether carbon emissions are really such a big deal. Does global warming really represent a significant threat to humankind? If so, what kind of threat? What will happen if we don't find a way to reduce emissions? Would a bit more warming necessarily be such a bad thing?

We'll take it as read that man-made emissions of carbon dioxide and other greenhouse gases cause global warming. Many people resist this idea, partly for psychological and political reasons that we'll explore later, but for now that doesn't warrant a diversion. Anyone who is in doubt should take note that every science academy in the world – without exception – agrees that the planet has been rapidly getting hotter and that emissions from human activity are almost certainly the main cause.[1]

Here, extremely briefly, are the core facts. Basic physics, known way back in the nineteenth century, tells us that additional greenhouse gases in the air will warm the planet; we've been releasing those gases and they've been measurably accumulating in the air; the Earth is responding just as the physics would predict; and there's no other plausible explanation for that warming.

Lest there be any doubt that the world is getting hotter, as we write this, January 2013 has just been announced by the US government's National Climate Data Center as the 335th consecutive month to have been warmer at the global level than the twentieth-century average for that month.[2] The last time a month fell below the century average was February 1985, around the time when Ronald Regan begun his second term in office and Mikhail Gorbachev became General Secretary of the Soviet Communist Party. Some people argue that the temperature records must be faulty, but there's a huge wealth of corroborating evidence, such as melting glaciers and sea ice, shifting growing seasons and wildlife, rising sea levels and greater humidity.[3]

As climate sceptics often (and rightly) point out, the world's climate has changed before as a result of natural processes and cycles. Tens of thousands of years ago, huge ice sheets enveloped much of Europe, Asia and North America, while in the planet's earliest days, the temperature was so high that all the water existed as a gas. It didn't cool enough for oceans to form for around a billion years. The climate will unavoidably change in the future, too. Within a couple of billion years, the Sun will grow and heat up sufficiently to move its 'Goldilocks zone' – the area with the right level of heat to support life – from Earth towards Mars. Five or six billion years after that, our star will grow into a red giant, absorbing the Earth in its fiery reach along the way. By that time the beings that inhabit the earth may be as different from us as we are from bacteria.[4] But the simple fact is that human civilisation has evolved and thrived in a short period of relative climatic stability. The concern is that we might be endangering that stability – and experts in the history of Earth's climate are among those raising the alarm.

No one is suggesting that climate change science is perfectly understood, however. On the contrary, there is uncertainty at every turn, as just about every scientist working in the area is keen to acknowledge. We don't know precisely how much warming our future emissions will cause. We don't know exactly how the climate will respond to that warming, nor how well human society and the world's wildlife will be able to cope. And we don't know if or when the climate system will reach a tipping point at which it switches into an altogether different state from the one that has allowed human society to evolve. With so many unknowns, it's tempting to shut our eyes and hope for the best as the world has been doing for the last two decades. A more sensible approach would be to take stock of what's at stake, review the best available evidence and choose how much risk we're prepared to take.

Climate impacts: the good, the bad and the beast

So far, a century or two after fossil fuel use took off during the industrial revolution, we've warmed the planet by around three quarters of a degree Celsius.[5] Although the warming has been higher over the land, for most of us as individuals, it has been barely perceptible and – given that we're used to temperature swings dozens of times larger within the space of a day – it's not immediately obvious that it's such a big deal. Indeed, those of us in cool countries spend most of our time longing for warmer weather. The cold can be both uncomfortable and expensive, forcing us to crank up the heating for half the year to keep warm. Hence some economists argue that a small amount of global warming is a good thing, having calculated that the benefits in cool countries – including fewer people dying in winter – outweigh the risks and inconvenience to those in hotter regions.[6] There may also be some other benefits of a small rise in temperatures. Some research has suggested that global agricultural productivity could increase a fraction thanks to longer growing seasons in temperate areas plus the fertilising effect of more carbon dioxide in the air.[7]

Even more remarkably, it's possible that by increasing the temperature a little we've deferred the next ice age. A recent study published in the journal *Nature Geoscience* examined previous interglacial periods and concluded that with low carbon dioxide concentrations, the next ice age could have begun at any time, but that thanks to the elevated concentration caused by human activity, 'No glacial inception is projected to occur'.[8]

Worryingly, though, even the small temperature rise we've seen so far is having some large-scale effects. The changes have been especially noticeable in the Arctic, where the temperature rise has been greatest and summer ice cover has collapsed much faster than most experts anticipated – to around half the level observed just a few decades ago. But the effects are also being felt across the wider world. A growing body of evidence has demonstrated that rising temperatures have significantly increased the likelihood of various types of extreme weather events. Recent analyses – some using models and others looking at historical records – have shown that devastating incidents such as the European heat wave of 2003, which killed tens of thousands, the Russian wildfires of 2010 and the Texan drought of 2011 were all made *many* times more likely by our carbon emissions.[9] (Oxford professor Myles Allen likes to talk about 'loading the climate dice' and points out that if you double the odds of rolling a six, which might represent a very rainy day, then you quadruple the odds of a double-six, which might represent a major flood.) Sea levels are also creeping up, and ecologists have highlighted how sensitive the world's ecosystems can be to relatively small changes in temperature. Just as seriously, much of the carbon dioxide we emit is getting absorbed by the ocean, rapidly driving up its acidity and posing profound risks to important marine life such as coral reefs.

None of these effects are trivial but if they were the end of the story we'd most likely learn to live with them. Only a handful of eco-warriors would be talking about reinventing our energy systems or rethinking economic priorities. Unfortunately, it's not the end of the story. It is the beginning.

We haven't yet seen the full effects of the carbon we've emitted so far. Because of the capacity of the oceans to store heat, our emissions take decades to reveal their full effects on the temperature. As a result of this planetary inertia, scientists estimate we've already locked in more than another half a degree of warming. That would be enough to take us to around 1.4°C above preindustrial levels, even if we switched off all our cars, factories and laptops tomorrow and never again burned a single lump of coal or drop of oil.[10] Any emissions we release from today – and remember the carbon curve is still accelerating upwards – will add to that.

As we look ahead into the future, the uncertainties grow. What would 1.4°C look like? How about 2°C? Or 4°C? Or 6°C? No one really knows the answers to these questions. But scores of scientists have spent decades examining and modelling past, current and future climates to make informed predictions. What's clear from their work is that there are two separate types of risk. First is the risk that any given temperature rise will bring specific problems – such as a foot of sea level rise or a 10 per cent increase in extreme downpours in a given region. This huge body of scientific research suggests that – on current emissions trends – we can expect to come up against all sorts of impacts of this type in coming decades. The list includes an increase in the number of devastating heat waves, droughts and floods; more water scarcity, more species loss and more intense hurricanes. (There will doubtless be plenty of less obvious effects too. To give just one of countless possible examples, a recent study predicts major problems for the incredibly climate-sensitive *arabica* plant beloved by the world's coffee drinkers.[11])

All this is cause for concern, but the second category of risk is considerably more worrying. That is the possibility that any given temperature increase would be enough to push the climate to a 'tipping point' where positive feedback loops kick in, flooding the air with greenhouse gases stored in soils and forests and accelerating the warming until the entire Earth system flips into a new state that might be dramatically less hospitable for humans. We know those feedbacks exist and we know some of them are already at work. For example, more and more carbon dioxide and methane is already leaking into the air from thawing

Arctic soils, and melting sea ice is decreasing the reflectivity of the planet's surfaces, causing more warmth from sunlight to be absorbed. We also know that the Earth's climate has undergone periods of rapid change before.

What we don't know is how hot the world would need to get before these feedbacks collectively run away with themselves, removing climate change from human control and committing the planet to irreversible future warming. Campaigners usually describe this kind of scenario as 'runaway climate change' or 'catastrophic global warming'. Scientists sometimes prefer more prosaic terms such as 'large-scale discontinuities', though legendary oceanographer Wallace Broecker perhaps captures the situation best with his image of a 'climate beast' that has been asleep long enough for human civilisation to thrive but is now being 'poked with a sharp stick'.[12] When will the beast wake up? No one can say for sure. What will happen when it wakes? Again, we can't know precisely, but it could very plausibly lead to a collapse in global food production and the extinction of many or even most of the world's plant and animal species. Large land areas could be lost under water, followed by many of the world's great cities. How long would this take? We don't know. Would the worst happen? We can't be sure.

Given all these uncertainties, dealing with climate change is ultimately all about juggling risk. Deciding how much hotter we allow the world to get before doing something about it means deciding how many negative impacts we're prepared, as a world, to sustain and how much chance we'll allow of a runaway climate disaster. How we juggle those risks depends, of course, on many other factors, such as how difficult and expensive we believe it will be to cut out fossil fuels; how much we value the future compared to the present; and how much hope we hold out that future generations will come up with currently implausible technical fixes to solve the problem if we fail.

How hot is too hot?

Having spent twenty years weighing up the risks and considering the relative costs of action and inaction, world leaders finally agreed at the UN climate talks in Copenhagen in 2009 that a temperature rise of two-degree Celsius above preindustrial levels was the line over which global society should not step.[13] As we'll see later, the policymakers who agreed this goal have so far stopped short of doing anything to ensure that it is going to be met, but at least they set a target. Two degrees was never universally accepted: low-lying island nations claimed it was insufficiently ambitious, pointing out that unless warming is limited to around 1.5°C their territories will most likely be submerged, even if climate change doesn't run away with itself. But the 'guardrail' was set at 2°C on the grounds that this level of temperature rise would most likely bring manageable – if still serious – impacts and keep the chance of runaway climate change acceptably low. The assumptions were informed by risk assessments published by the UN's Intergovernmental Panel on Climate Change (IPCC) back in 2001.

With unfortunate timing, in the same year as Copenhagen, a team of scientists published a review and update of the IPCC's risk assessments. These were dramatically more severe, suggesting that an increase of just *one* degree would be about as risky as two degrees was thought to be back in 2001 and that in terms of runaway warming, two degrees was about as dangerous as 4°C had previously been thought to be.[14] Veteran climatologist Jim Hansen of NASA has gone so far as to describe the idea of allowing the world to get to 2°C as a 'prescription for disaster', arguing that the target is simply too lax to protect us from tipping points.[15]

These and other recent developments in climate science – such as a 2012 paper that showed that the more worrying climate models have proved the most accurate at predicting humidity changes[16] – lead to a pair of uncomfortable realisations. First, it looks like the warming we're already committed to is in itself likely to lead to some very serious impacts on ecosystems and people. Second, a 2°C target appears to be much more dangerous than was previously thought, both from the perspective of specific 'predictable' risks and, even more pertinently, in terms of the chance of crossing a climatic tipping point.

Beyond 2°C, the picture becomes ever more bleak, although the specifics become hard to predict. By the time we get to 4°C, there are huge uncertainties in the models but the potential impacts are – to borrow a phrase from the usually conservative UK Met Office – 'frightening'. They include, by 2080, when today's young children will be roughly at retirement age, a large reduction in global crop yields, significant stresses on water availability and massive threats to the world's plant and animal species. Marine ecosystems would be 'fundamentally altered'. Almost all the permafrost (deep-frozen soil) in Northern Europe and Siberia would most likely have thawed, causing significant further volumes of additional carbon dixoide and methane into the air, and committing to many more decades or centuries of the same. This positive feedback would drive up the temperature yet further. No one knows how quickly it would happen or when the rise would stop.

In the context of day-to-day experience, four degrees may not sound like a great deal, but it's comparable to the temperature change observed since the ice age's Last Glacial Maximum (around 20,000 years ago), when the earth looked fundamentally different.[17] Moreover, temperature rises wouldn't be distributed evenly across the world. In land areas there would most likely be a 5–6°C rise on average. The Met Office predicts that the most scorching days in New York might be as much as 10–12°C hotter than they are now, and those in Central Europe 8°C hotter.[18] It's not at all clear how much of our current essential infrastructure would continue to function in such a situation. As Kevin Anderson of the Tyndall Centre for Climate Change Research, puts it: 'There is a widespread view that a 4°C future is incompatible with an organised global community, is likely to be beyond "adaptation", is devastating to the majority of ecosystems and has a high probability of not being stable (i.e. 4°C would be an interim temperature on the way to a much higher equilibrium level).'[19]

Not everyone goes as far as Anderson, but even conservative bodies such as the International Energy Agency (IEA) and World Bank are clear about the direction of the science. The IEA *World Energy Outlook 2011* concludes: 'The uncomfortable message from the scientific community is that although the difficulty of achieving [a 2°C temperature limit] is increasing sharply with every passing year, so too are the predicted consequences of failing to do so.' In 2012, the organisation's director, Maria van der Hoeven, said that on current emissions trends we may be heading for a 6°C rise this century.[20] A few months later, the World Bank published a report saying that without 'serious policy changes' the world is on track for 4°C, a scenario its president Jim Yong Kim described as 'devastating'.[21]

In short, we need to face it that the world's 2°C climate target is much more than we should risk but much less than we are heading for.

3. The trillion-tonne limit

Why we need to stay within an all-time carbon budget, most of which has already been used

For a temperature target to be any practical use it needs to be translated into something we can influence directly: emissions. This adds yet another layer of uncertainty because we can't be sure precisely how sensitive the atmosphere is to the carbon we emit. But scientists have spent decades exploring the various factors in models to estimate the likelihood of different outcomes in different emissions scenarios. As the models have improved, something counterintuitive has become clear. Because carbon dioxide stays in the air for such a long time, and because of the way the planet responds, the eventual temperature the Earth will reach does not depend much on *when* we emit the carbon – just on how much we release in total, cumulatively. In other words, to have a good chance of avoiding any particular temperature rise, we need to remain within an all-time global 'carbon budget'.

The numbers, while necessarily very approximate, are rather neat according to a landmark study from 2009.[1] This suggested that if we are content with a fifty per cent chance of exceeding 2°C, we can emit no more than a total of one trillion tonnes of carbon, which equates to 3,700 billion tonnes of carbon dioxide.[2] We've already emitted more than half of that since the industrial revolution, which leaves us with a budget for the rest of time of around 1,600 billion tonnes. We'll call that the 'coin flip' scenario, as it means accepting an odds-even chance of crossing the globally accepted danger threshold. For a 75 per cent chance of avoiding that threshold – which would be more prudent given what's at stake, and how risky 2°C now looks – our remaining budget shrinks by more than half to around 700 billion tonnes.

We'll call this the 'double-coin flip' scenario, as the risk of failure is the same as two coins both landing on tails. To improve the odds further, the budget would need to shrink even more. (Hence Bill McKibben's number, in the foreword to this book, of 565 billion tonnes for an 80 per cent chance, which is based on another study that reached very similar conclusions.)

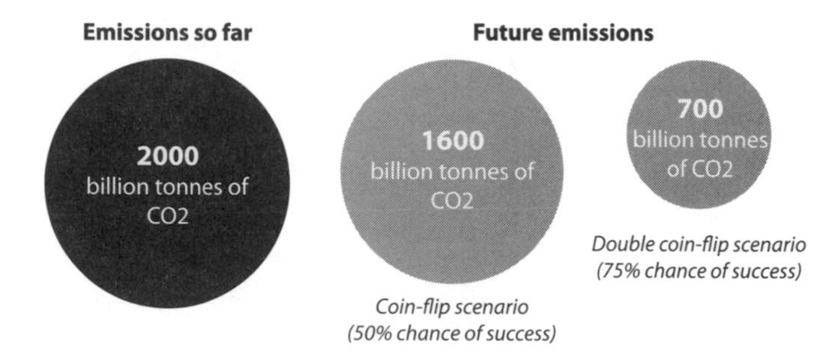

Carbon budgets based on the trillion tonne analysis. Some other studies, backed up by recent developments in estimates of climate sensitivity, suggest slightly larger budgets for any given odds.

These budgets are somewhat oversimplified as they consider carbon dioxide only, ignoring the other planet-warming gases and particles released by human activity. As we show in part four of the book, dealing with these is essential for limiting global warming – and especially for holding down the temperature in the next few decades – but doing so can't substitute for limiting cumulative carbon dioxide emissions from fossil fuels.

Putting the task in context

Combining these budgets with the long-term carbon curve gives a sense of the scale of the task ahead. Readers who paid attention to maths in school will know that the area under the curve represents the total volume of carbon dioxide released – and is therefore what we need to constrain. The graph above

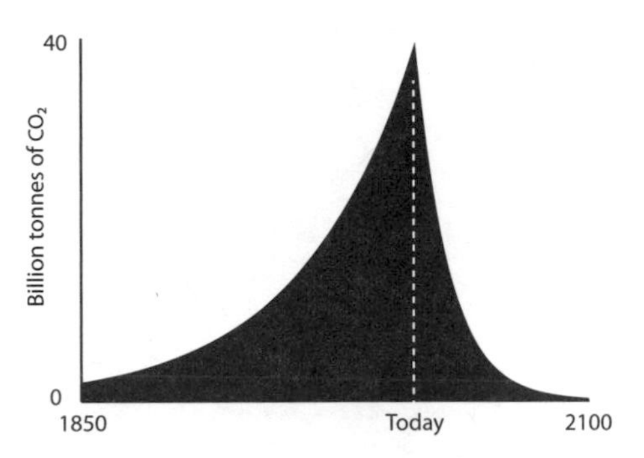

Global carbon dioxide cuts required for a 75 per cent chance of not exceeding 2°C, with a peak today

shows the kind of turnaround that would be needed to provide a 75 per cent chance of keeping within 2°C if we started reducing emissions at a steady rate today.

In reality, we'd never achieve that kind of alpine peak on the curve, because such an abrupt kink would likely inflict mayhem on the global economy. A more plausible pathway for success would see the curve continuing upwards but slowing down for a few years before levelling off and falling. But each year we delay before reaching the peak, the faster the subsequent cuts will need to be if we're to stay within any given budget. As the graph overleaf shows, if we leave it until 2020 to reduce global emissions – which is the absolute best that the international negotiations are aiming for – then a very rapid turnaround is needed even if we accepted coin-flip odds. The cuts required after the peak would be around 3 to 4 per cent a year for decades to come. This is a very ambitious rate rarely seen so far except during times of recession. Moreover, a *much* greater rate of decline – around 10 per cent a year – would be needed in rich countries if the developing world reasonably insisted on capturing the majority of the remaining budget to reflect its larger population, greater poverty levels and relatively low level of historical emissions. The cuts would need to be even steeper still if deforestation isn't

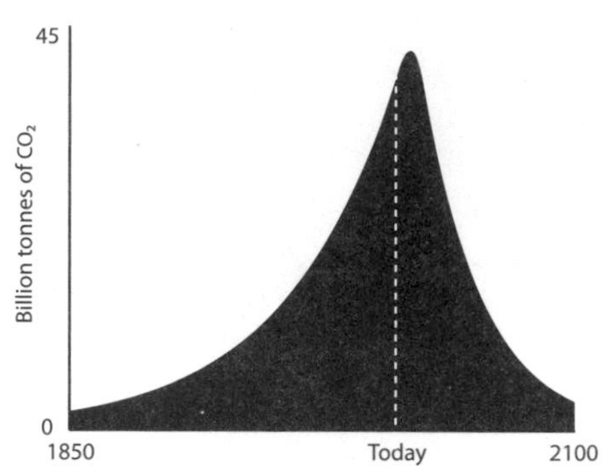

Global carbon dioxide cuts required for a 50 per cent chance of not exceeding 2°C, with a peak in 2020

rapidly stopped – or if more carbon than expected leaks into the air from melting permafrost.

As we'll show later, it's possible to construct a scenario in which cuts of the required speed are made without a major impact on global affluence. With the right policies in place, new energy technologies could potentially scale up fast enough to replace rapidly diminishing flows of fossil fuels. But so far, that's not happening. Instead, the carbon curve is continuing up its long-term path and the 2°C target is looking tougher every year – to the point that many people working on climate change now say, either on or off the record, they no longer think it's possible. Even accountancy giant PricewaterhouseCoopers, which tracks the world's carbon emissions relative to economic activity, stated boldly in 2012 that 'ambitions to limit warming to 2°C appear highly unrealistic.'

Clearly the world faces a huge challenge. We'll explore the nature of that challenge in parts two and three, but there are a couple of important questions about our current situation that we first need to address. If fossil fuels are the root cause of the problem, might their growing scarcity help level off the curve? With all the talk of peak oil and dwindling resources, could rising fuel costs and shrinking reserves come to the rescue?

4. Too much fuel in the ground

There's far more oil, coal and gas than we can safely burn

The climate change debate tends to focus on emissions – the carbon dioxide that rises from our chimneys, tailpipes and flues. That's not surprising, because the greenhouse gas pouring into the atmosphere is the direct cause of the warming. But everything that gets burned has to be taken out of the ground first. To understand the problem properly, it's necessary to look at the fuel reserves, and to compare their carbon content with the emissions budgets from the last chapter. When you do, a key fact emerges: for all the talk about finite resources and peak oil, scarcity is resoundingly not the problem. From the climate's perspective, there is far too much fossil fuel. The problem, in fact, is abundance.

No one knows precisely how much fossil fuel exists. Even if they did, the figure wouldn't mean much because not all fuel deposits can be viably extracted at any given time. Some oil wells gush out liquid fuel when drilled and many coal seams exist close to the surface, enabling them to be easily and inexpensively mined. By contrast, a significant slice of the world's remaining oil exists in thick, gloopy forms – such as bituminous tar sands – that take lots of energy, effort and infrastructure to turn into useful fuels. Similarly, a huge volume of natural gas is trapped in shale rocks that can only be released by a process known as hydraulically fracturing – 'fracking' – which involves pumping pressurised water, sand and chemicals into the rock.

The amount of fossil fuel that can be viably extracted and processed therefore depends on at least three other factors. First, the current technology and know-how of fossil fuel companies, which determines the cost of extraction. Second, how much people and companies are prepared to pay for energy. Third, the regulatory regime, including where oil and gas drilling is permitted, what level of royalties are demanded by the host government, if any, and any taxes or subsidies that raise or lower the price of using the fuels. These factors change over time, and the amount of commercially available fossil fuel varies with them. For example, shale gas remained largely irrelevant until horizontal drilling technology reached maturity and gas prices were high enough to enable companies to operate fracking sites at a profit. Today, however, in the US at least, fracked natural gas is abundant and cheap. Yet another factor is the physical state of the planet: melting ice caused by climate change is opening up whole new areas of the Arctic to oil and gas exploration.

With so many variables, any estimate of the total amount of fossil fuel we could eventually burn involves plenty of guesswork. Hence energy experts differentiate oil, coal and gas *resources* – the estimated deposits in the ground – from *reserves*, which describes the share of those resources which are economically and technically viable and have a statistical probability of being produced. Reflecting the uncertainties involved, reserves are usually broken down into *proven reserves* (an estimate with a 90 per cent chance of the stated number being exceeded), *probable reserves* (with a 50 per cent or more chance) and *possible reserves* (with a 10 per cent chance or better).[1]

Much of the easiest fuel to access and process has already been burned, but every year the industry discovers new pockets and gets better at extracting the reserves that are hard or more expensive to reach or refine. Hence, over time, undiscovered fuels expand the known resource base and inaccessible resources become possible, probable or proven reserves. Sometimes this process happens at an alarming rate. When we first drafted this chapter, the proven oil reserves listed in the standard industry reference were 1383 billion barrels. By the time we came to the second draft, the total had increased to 1653 billion – a remarkable increase mainly reflecting unconventional oils from Canada and Venezuela being reclassified as proven reserves and added to the official books in significant volumes for the first time.[2]

Carbon's journey from hidden underground to invisible in the air

It should be said that estimates for both reserves and resources need to be taken with a pinch of salt. One reason is that owners of fossil fuel deposits usually have a financial interest in their assets being listed as larger rather than smaller. This applies both to companies – as evidenced by Shell's high-profile exaggeration of its proven reserves in 2004[3] – and to countries, some of which are widely believed to have overstated their reserves to increase their share of oil exports within the OPEC cartel. (The imaginary fuel stocks that may have been added to their books in the 1980s are sometimes dubbed 'political reserves'.) Another reason for scepticim, in the case of resources, is the sheer degree of uncertainty. Estimates for technically recoverable shale gas deposits in large countries have sometimes literally doubled or halved in the space of a year.[4]

When it comes to climate change, though, none of these details matter much because even if we focus just on the proven reserves, it's clear that the world has far more fossil fuel than it can safely burn. The chart overleaf shows this by putting the 'coin-flip' and 'double coin-flip' carbon budgets from the previous chapter into the context of the standard figures for proven reserves. Very roughly speaking, we can burn half of them if we are prepared to accept 50 per cent odds of staying below 2°C; or just a quarter of the reserves if we want 75 per cent odds.

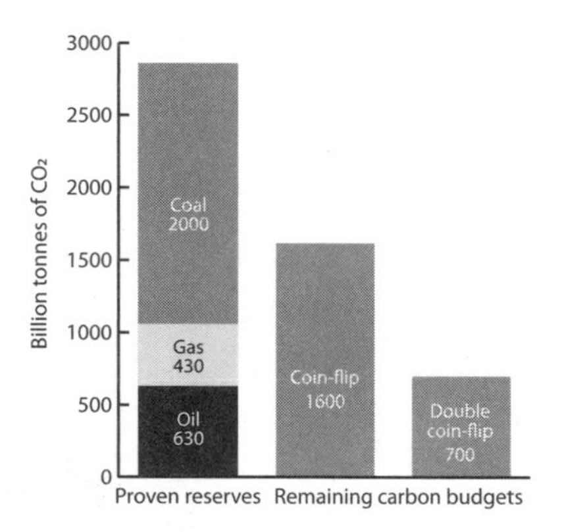

Potential carbon dioxide emissions from proven (economically recoverable) fossil fuel reserves compared to carbon budgets for a 50 and 75 per cent chance of limiting global warming to 2°C.[5]

When you consider how much fossil fuels are worth and how much we've already invested in machines to burn them, this chart looks daunting. Looking on the bright side, if we accepted coin-flip odds we could in theory burn all the proven oil and gas reserves – but only if we gave up on all the other reserves and resources, immediately halted deforestation and other sources of carbon dioxide *and* quickly reduced coal consumption to almost nothing, or at least fitted out all coal power plants with equipment for capturing the carbon.[6] That is not the direction in which the world is moving, however. As we'll see later, unabated coal burning has been soaring in recent years. More concerning still, the chart above only considers proven reserves, ignoring the much larger probable reserves and almost all the unconventional sources such as tar sands and oil shale which energy companies are quickly bringing on stream.

If we factor in all those additional resources, the fossil fuel stack gets a whole lot bigger. At this point any estimate contains a significant amount of uncertainty, but it's still possible to get a sense of scale. Research by Christophe McGlade of the Energy Institute at University College, London, suggests that once all the tar sands, heavy oil, kerogen, shale gas, coal-bed methane and other unconventional sources are included, the world may have over five trillion barrels of recoverable oil and 680 trillion cubic metres of recoverable natural gas.[7] These figures roughly match the most recent estimates from the IEA, which calculated that the total recoverable oil and gas resource is around 3.5 times greater than the proven reserves alone.[8] The recoverable coal resource is even bigger still: around *twenty* times the proven reserves. If we update the graph to reflect all this, things look radically different.

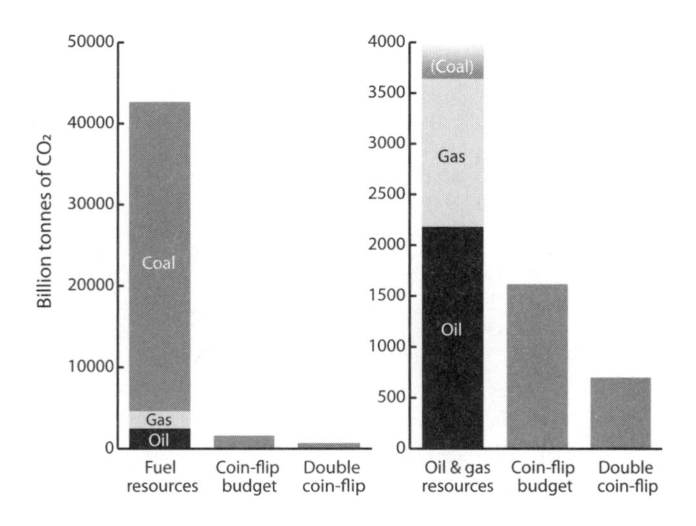

Left: potential carbon dioxide emissions from recoverable fossil fuel resources compared to carbon budgets to provide a 50 or 75 per cent chance of avoiding 2°C.[9] Right: the same graph shown without coal for ease of viewing.

The picture is now very clear. We already have *far* more fossil fuel reserves than we can burn in the usual way, with the carbon dioxide flowing into the air. And there are plenty of additional resources that could top up those reserves over time. The inescapable conclusion is that to tackle global warming, we need to leave most of the world's fossil fuel reserves in the ground. That's going to be tough because they're immensely valuable, both to their owners and to the wider global economy. We'll explore the difficulty of leaving reserves intact later, along with the possibility that we could keep burning the fossil fuels but capture and bury the carbon. We will also look at the possibility of trying to render the fuels worthless by making alternative energy sources so cheap that the oil, coal and gas aren't worth extracting. For now, though, just hold the thought that there's enough fossil fuel left to drive us far into a zone of extreme danger.

Peak fossil fuels

At first glance this analysis clashes with many of the arguments made over the years about 'peak oil' – the name given to the point when the world's production of oil reaches its final peak and begins an inexorable fall. Given that oil is a finite resource, there's no question that a peak will arrive, but there's an active debate about its timing and implications. Analysts and commentators concerned about peak oil – the so-called 'peakists' – have long argued that this moment is imminent or perhaps even already in the past. They claim that when markets wake up to the fact that oil production has begun to decline, prices will skyrocket and global economic meltdown may ensue.[10]

In the last few years, it has looked increasingly plausible that *conventional* oil production – in the form of crude oil and natural gas liquids – will indeed peak soon. In its 2012 *World Energy Outlook*, the IEA predicted that output would roughly level off by 2015.[11] This is a vindication of sorts for the peakists, but the agency also predicts that conventional oil will plateau for decades, rather than falling rapidly and taking the economy with it. More importantly, the IEA expects *total* production of liquid fossil fuels to keep growing even after crude has peaked, thanks to increasing supplies of unconventional oils made from tar sands, kerogen, coal and other sources.

Whether or not that happens, one thing that's clear is this: even if peak oil is reached soon, there will still be plenty enough oil, coal and gas to play havoc with the planet's climate. An early peak could potentially scare governments into investing more in renewables and nuclear; or it could cause a global depression, slowing the rate of emissions growth for a while. But it won't persuade the owners of oil reserves to leave most of the remaining crude in the ground. And even if it did, we'd still have coal and gas to worry about. Indeed, for all the talk of peak oil, solving climate change needs something much bigger. Barring enormous breakthroughs in carbon capture technology, stopping global warming means humankind deliberately bringing about *peak fossil fuels*. And not just a peak, but a sky-diving decline.

5. No deal on the horizon

The world's governments are not remotely on track to meet their own target

As we've seen, emissions are continuing to shoot up, even as the science is getting scarier. We are eating through our all-time carbon budget at terrifying and accelerating speed. And the world has far more oil, coal and gas than we can safely burn, at least until technology exists to put the carbon back into the ground. The situation is stark, so how is the world responding? Are policymakers, innovators, businesses and concerned citizens preparing the ground for a solution? Or is the world sleepwalking towards a cliff edge?

Responses to climate change broadly follow two paths. There's the top-down approach, where nations negotiate to limit global emissions in an orchestrated way. Then there's the bottom-up approach, in which concerned families, communities, companies, states and regions press ahead of the crowd and try to limit their own emissions, motivated either by wanting to do the right thing or by the desire to save money on energy. There has been effort put into both approaches in the last decade or so.

On the bottom-up front, there's a buzz of activity on every level. Almost every major business now has someone whose job is to monitor and try to reduce carbon emissions. Public information posters, local government newsletters and even utility bills encourage us to save energy at home and on the road. Environment groups have joined forces with other concerned organisations and corporations to call for change. National governments incentivise renewable energy production and many companies strive to make more efficient cars and appliances. A small but by no means invisible minority of people are proactively limiting their carbon footprints by, for example, flying less frequently or changing their buying habits.

The next section of the book shows why these bottom-up responses, crucial though they are, haven't so far dinted the carbon curve. This chapter focuses on the other part of the picture: the world's efforts to negotiate a top-down solution. According to surveys, most of us feel intuitively that this is the natural way for a global issue such as climate change to be solved.[1] After all, if the risks are so big, wouldn't world leaders be putting a solution in place on our behalf?

The story so far

Serious concern within the scientific community about climate change started stirring in the 1970s and 1980s but politicians didn't really take note until 1990, when the Intergovernmental Panel on Climate Change (IPCC) – a special UN body put together to summarise the science of global warming – published its first major report. Two years later, at a landmark environment summit in Rio de Janeiro, representatives from more than a hundred countries signed a treaty designed to 'prevent dangerous anthropogenic interference with the climate'.[2] As part of this, countries agreed to a few simple measures such as starting to keep track of their greenhouse gas emissions, but many nations fiercely resisted any obligation to actually reduce those emissions.

Five years and many meetings later – in Kyoto, Japan – the world finally agreed a deal with legally binding targets for emissions cuts. Only rich countries were covered, however, and the targets weren't very ambitious. The average required cut was a 5 per cent reduction by 2008–2012, relative to 1990 levels. As many campaigners were quick to point out, that was never going to be enough to solve the problem, but the Kyoto Protocol did represent an important first step in global climate diplomacy.

Unfortunately, the chance of that first step leading painlessly to a second was significantly reduced when the world's then-biggest emitter – the US – decided not to ratify the treaty. In other words, America signed up to the agreement but refused to implement it. The obstacle was a motion in the Senate called the Byrd–Hagel Resolution, which passed by a vote of 95–0. It stated that the US wouldn't commit to binding emissions cuts unless developing countries were also obliged to do so.[3]

The rest of the developed world pressed on regardless and ratified the protocol. Russia and especially Australia both wobbled for years but eventually came on board. Canada, on the other hand, ratified but then pulled out in 2011 – towards the end of the Kyoto period – arguing that it was going to miss its targets and wasn't prepared to pay the associated costs of doing so. (Its position emphasised quite how delicate even hard-won international agreements can be, a point we'll come back to later.) Although the global carbon curve doesn't appear to have even noticed the Kyoto Protocol, in the meantime discussions have continued to try to reach a more ambitious and genuinely global deal to follow on from it.

With this aim, hundreds of delegates – not to mention thousands of campaigners, journalists and lobbyists – assemble in a different city each year to try to hammer out an agreement. At the time of writing, there have been *eighteen* formal summits, everywhere from Bali to Cancun, with countless pre-meetings in the run-up to each. The mainstream media generally pays little heed to this endless diplomatic circus but that changed briefly in 2009 when interest was heightened by a sense of optimism surrounding the meeting in Copenhagen. In the event, the optimism turned out to be misplaced. Copenhagen did see a breakthrough of sorts: an agreement in principle between the world's nations to try to limit global warming to 2°C, the temperature target discussed in chapter two. But when it came to emissions cuts, all the negotiators could manage was a voluntary regime

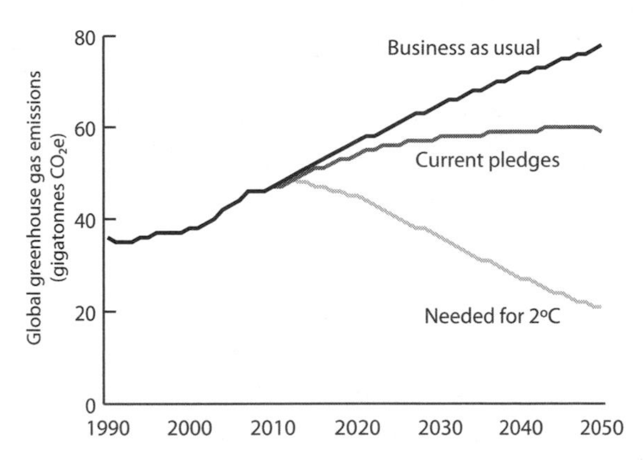

The gap between carbon cutting pledges made so far and the cuts needed for a good chance of limiting global warming to 2°C, based on data from ClimateActionTracker.org

whereby each country would publish non-binding pledges for the period up until 2020.

Some nations, most notably those in Europe, have made significant pledges to cut emissions. Taken together, though, these pledges are a small fraction of what is required to keep danger at bay. A group of scientists from various European organisations have estimated the current round of pledges might be sufficient to slow down and perhaps level off the global carbon curve by around 2050, the most likely result being an extremely dangerous global temperature rise of 3.3°C. But even that scenario (which is shown in the graph above) is optimistic, because – as we'll argue in the next part of the book – not using fossil fuels in one region isn't enough to keep them in the ground. They have a tendency to get burned elsewhere. There's also the possibility that voluntary national targets may not be hit. As Canada has shown, a pledge isn't a guarantee of success. Moreover, some of the current national pledges are commitments to slow the growth of emissions relative to economic activity rather than actually cut them. For example, China has pledged 'to endeavour to lower carbon dioxide emissions per unit of GDP by 40–45 per cent by 2020'. Hence even if the target is met, the actual level of emissions will depend on what happens to the country's GDP.[4]

It's even possible that by promising to cut emissions but not actually doing so, world leaders have encouraged the fossil fuel sector to increase production in order to cash in their reserves while they still can – a concept dubbed the 'green paradox'.[5]

Things *have* progressed a little since Copenhagen, however. At the end of 2011, the world's nations agreed to a roadmap that promises to deliver a legally binding global deal on carbon emissions. This was widely reported as a breakthrough, and in a limited sense it was. For the first time, emerging economies such as India, Brazil and China agreed that they would accept legally binding commitments on climate change. Unfortunately, there were two important catches. First, the timeline: although the deal must be agreed by 2015, it won't come into force until 2020. That's bad news, because – as we saw in chapter two – every year in which emissions keep rising will make the task that follows harder. The second catch is a conspicuous absence of any numbers. Rather than the sequence of events that would seem most logical – starting with the agreed temperature target of 2°C, converting that into a global carbon budget and then negotiating who gets what – world leaders have agreed only to agree *something*. So while the national-level targets that eventually get pinned down may be legally binding, there's no promise that they will be enough to add up to success. In fact, deliberately or not, the roadmap was worded in such a way that nations could in principle legally oblige themselves to *increase* rather than decrease their emissions.[6]

The wrangles

Given what's at stake, why can't governments just agree a deal ambitious enough to solve the problem? There are many answers to this question, from the relative powerlessness of the negotiators (who are typically energy and environment ministers rather than presidents and prime ministers) through to the nervousness of many governments to do anything that may create political tension at home, such as an increase in fuel prices or a slowdown in GDP. There's also the all-important issue of vested interests. Even if the whole developing world *did* agree to a deal, many American policymakers would continue to block US participation on the grounds that climate change isn't a serious enough threat to make it worth endangering the national economy. It doesn't take a private detective to spot that many of these politicians rely heavily on the energy sector for their campaign funding.

We'll return to these themes later in the book, but in the meantime let's zoom in briefly on the most pervasive obstacle of all: the question of which nations should have to make what

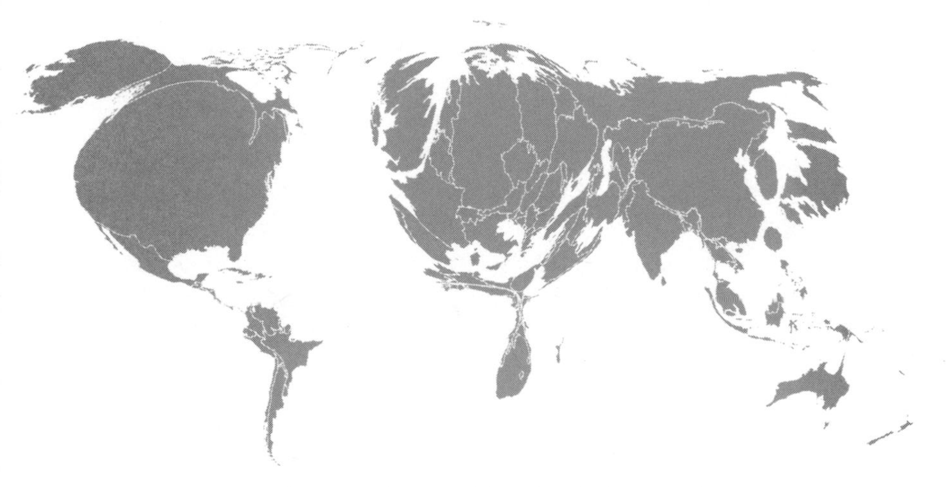

A map showing countries resized to reflect their cumulative carbon emission from energy use, 1850–2008.[7] The US, the UK and Germany are all bulging, while Africa and South America are almost unrecognisably small.

levels of sacrifice. This is the fracture that runs consistently through the negotiations – the justification used by nations from America to India to resist proposals. One important factor here is the question of historical responsibility. Because energy and economic activity are so closely bound together – as we'll see later in the book – the richest countries tend to have burned far more fossil fuels per person in previous decades and centuries than poorer nations did. In other words, developed countries have caused most of the global warming experienced so far. And because carbon dioxide can stay in the air for centuries, those historical emissions will continue to cause warming for a long time to come. Europe and the US alone have caused around half the carbon emissions since 1850, despite having a much smaller slice of the total population.

The developed world burned all those fuels without solid knowledge about global warming, so it's more a case of accidental damage than deliberate harm – at least until the last twenty years or so. Nonetheless, those of us who live in relatively rich countries benefit today from infrastructure, industries and economic advantage developed with the aid of copious fossil fuel supplies. Now there's a need to constrain fuel use globally, it's no wonder that those countries which haven't yet developed to the same degree are reluctant to make commitments until the wealthy world with its greater historical responsibility moves first – especially given a set of broader historical themes such as slavery and colonialism, all of which feed into the political picture. Moreover, rich countries – by virtue of being richer – are in a better position to develop and fund the new technologies needed to replace fossil fuels.

On the other hand, it's no longer the case that 'developed' countries are responsible for the majority of the current carbon emissions. When Kyoto was negotiated in 1997, the US was by far the world's biggest emitter and the rich world as a whole accounted for most of the global carbon footprint.[8] But China

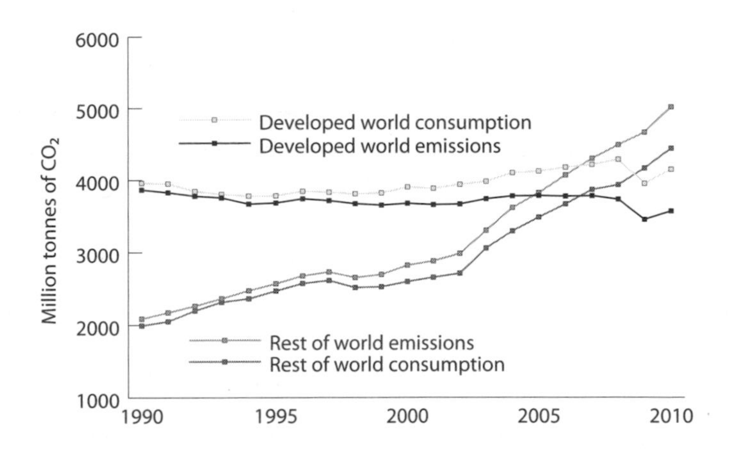

Rich countries, as defined by the Kyoto Protocol, now produce a minority of the carbon emissions, even if the consumption of imported goods is taken into account.

overtook the US as the leading emitter in the early years of the new millennium and today OECD countries release only around a *third* of the carbon dioxide – a much smaller proportion than most experts would have predicted a decade or two ago. Even when the carbon footprint of imported and exported goods is taken into account, rich countries account for less than half of the total. So even if they promised to cut their emissions to zero by 2050, that still probably wouldn't be enough to solve the problem. Swiftly rising emissions from China, India, Brazil, South Africa and other emerging economies would on current trends easily exceed the global carbon budgets described in chapter two.

In the logjam that results from all this, some rich countries – such as the US, Japan and Canada – are saying they won't commit to ambitious cuts unless emerging economies also make meaningful pledges. In turn, the emerging economies are saying that they won't make ambitious commitments until the rich countries show more leadership. At first glance, it all sounds like political posturing but to really understand what's going on here, it's necessary to look back at the charts in the previous chapter. If the negotiators opted for a global carbon budget in line with their own temperature target, that would mean leaving most of the world's fossil fuel reserves plus all the unconventional resources in the ground, at least until carbon capture technology is cheaply available. In other words, they'd be accepting that the world will burn only a slice of the remaining fossil energy reserves and trying to agree how to divide that slice between them.

Seen in these stark terms it's perhaps less surprising that the conversation has proved so difficult – and it becomes obvious that to get a full picture we need to think about which nations actually own all the fossil fuel reserves that might need to be written off. We'll come back to this question in part three. But first, let's shift from the top-down global politics to the bottom-up efforts of concerned citizens, companies and governments that have sought to reduce fossil fuel use with energy-efficient technologies, greener behaviours and clean energy sources. Many such efforts have been made, and social trends and government policies have also helped to slow population growth. But so far none of this appears to have made any difference to the global carbon curve. It's as if reducing fossil fuel use is like squeezing a balloon: cuts made in one place reappear as additional use elsewhere. But why would that happen? And what does it mean?

PART TWO

Squeezing a balloon

Why many of the things we think should help get cancelled out at the global system level

6. Rebounds and ripples

How efficiency gains and local savings bounce back

In discussions about cutting carbon emissions, the word 'efficiency' crops up a lot. Energy-efficient homes and cars. Energy-efficient power stations. Energy-efficient processes and factories, behaviours and lifestyle choices. Energy-efficient economies. The prevalent assumption, understandably enough, is that if we use energy more efficiently, the result will be less energy use overall. If our cars get twice as many miles to the gallon, surely we'll burn half as much fuel. And if a company 'saves a tonne of carbon dioxide' though energy-efficiency improvements, then simple logic would suggest that a tonne less carbon dioxide will flow into the air.

But that is not how it turns out. This chapter explores the counterintuitive ways in which the global economy absorbs efficiency improvements and local energy savings. Although, for reasons we'll come to later, efficiency and voluntary carbon cuts do have an essential role to play in bringing about a low-carbon world, it turns out that on closer inspection, they don't in themselves do much, if anything at all, to cut global emissions.

Why energy efficiency isn't all it seems to be

Over the centuries, the world's engineers and scientists have proved endlessly ingenious at reducing the amount of energy needed to achieve a particular outcome. Some modern cars can go more than four times as far on a gallon of petrol than a Model-T Ford did a hundred years ago, despite also going more than twice as fast.[1] We can store information digitally billions of times more efficiently than filing cabinets allowed us to do. The energy it takes to move a tonne of freight by road or sea, or to heat a new home, has collapsed since 1900. A modern LED throws out light more than a hundred times more efficiently than a candle and around five times more efficiently than a traditional tungsten filament bulb. In fact, with a few notable exceptions – such as the craze for large four-wheel-drive cars in the US – energy efficiency of almost everything we do has been improving in leaps and bounds.

These gains have enabled us to get richer by extracting more usefulness from each unit of available energy. Indeed, as we'll see in the next chapter, the energy we use to generate each unit of economic activity – each dollar, pound or euro earned and spent – has been falling fairly steadily for decades, even centuries. All of this is great news. But that trend has gone hand-in-hand with another: the inexorable rise in both energy use and emissions that we described in chapter one.

It's tempting to try and resolve this apparent paradox by assuming that the problem with energy efficiency is that we haven't had enough of it. If society had tried harder and increased efficiency more, then perhaps we'd have achieved the same economic activity using less energy and generating fewer emissions. If so, increasing energy efficiency in the future would in itself help us level off the carbon curve. But there's another completely different way of looking at the effect of efficiency gains which may fit the historical facts better. When we improve energy efficiency, we make energy more productive, because each drop of oil or lump of coal can do more work. As a rule, making something more productive makes it more valuable and that in turn means we use more of it. Think of it this way: if it took a tonne of coal to keep a household warm for a night, or a barrel of oil to drive to the shops, we probably wouldn't bother with fossil fuel at all.

The other key point is that our machinery and technology for extracting fossil fuels have become more efficient over time. So, seen in the round, efficiency improvements have increased not only the demand for fuel but also the supply. Could it be, then, that they have led to *increased* total energy use?

A background to rebound

The debate about the impact of energy efficiency has been running for at least a century and a half. Writing in 1865, when coal-powered Britain was continuing its imperial expansion, William Stanley Jevons urged the country's leaders to consider what would happen to British influence when the coal failed to keep up with demand. In his book *The Coal Question*, he agued that using coal more efficiently wouldn't be enough to avoid the crunch, because better efficiency would just make the fuel even more attractive for use in ever more engines in ever more situations. Whatever 'conduces to increase the efficiency of coal and to diminish the cost of its use', he wrote, 'directly tends to augment the value of the steam-engine, and to enlarge the field of its operations.'[2]

A century and a half and plenty of academic analysis later, energy experts are still debating whether Jevons was right.[3] The argument boils down to what proportion of the expected energy savings of an efficiency improvement is lost through so called 'rebound effects': a metaphor describing the way in which savings from efficiency gains bounce back as additional energy use elsewhere. Some argue the rebound effect is small and that just a few per cent of the expected savings get lost. Others believe the effect can be much larger, even concurring with Jevons that the true figure is more than a hundred per cent: that efficiency gains tend to 'backfire' and drive up energy consumption overall. That certainly fits with the persistent historical trends of rising energy use and greater efficiency.

The idea that efficiency improvements could somehow fan the flames of climate change is disconcerting and stands to throw a good few green policies into confusion. How could insulating a building or making a car more efficient possibly be bad for carbon emissions? To answer that question we need to explore the dynamics of society's energy use more closely. We'll do that by looking at the various different ways in which rebounds can come about, from the most direct to the most diffuse.

Rebound type one: make it cheaper, use more

The most obvious rebound effect stems from the simple fact that when something becomes cheaper, we tend to consume more of it. The classic example is lighting. The journey from camp fires and candles to modern bulbs has seen the cost of illumination collapse – and the journey is still continuing, as incandescent bulbs give way to more efficient options such as compact fluorescents and LEDs. But these huge gains haven't slashed the energy we use for lighting. Instead, falling costs have enabled us to light more rooms at a time and more corners within each room. Sometimes this effect plays out in individual decision-making: once a household switches to efficient bulbs, its residents might be tempted to think 'I've left the lights on but its okay because they're low-energy ones'. In other cases, it's just a gradual shift in societal norms. Decorators and builders today often fit ten or more halogen or LED spotlights in individual rooms – a far cry from the 1980s when most rooms were lit by a single pendant light in the centre, plus perhaps a lamp or two. As a result, the total energy consumption for lighting hasn't dipped in line with efficiency gains. In fact, in the UK, the energy footprint of lighting increased by almost two-thirds between 1970 and 2000 – including an 11 per cent rise between 1990 and 2000 just when low-energy compact fluorescents started to rival the popularity of traditional tungsten filament bulbs.[4]

The IT revolution, which has transformed society but not noticeably impacted on the carbon curve, gives us another example. How can it be that the ability to store and send information digitally rather than by old-fashioned and inefficient filing cabinet, paper, train, plane and postal van hasn't delivered any observable cuts in energy use? A simple part of the answer is that as a result of the internet we store and send billions of times more data, whether that's sharing photos with friends on Facebook, sending large attachments to colleagues by email, or backing up our hard drives to online services. (As we write this book, each iteration of each file is being saved to Dropbox. It's handy but it means that our 70,000 words of text are using as much server space as perhaps a dozen large encyclopaedias might have taken up only a few years ago.) Another part of the explanation is that we don't swap one method for the other; we use them both. Most homes in the developed world now use email, but that hasn't stopped the daily visit from the post van. We download films to tablet computers but still have TVs.

This simple consumption-boosting rebound won't in most cases eradicate the carbon savings from improving efficiency. If as a household or a whole society we improve lighting efficiency by, say, a factor of five, we're unlikely to use up *all* the savings by having five times as much light – especially once we've reached the point where our ceilings are jam-packed with bulbs. But that's not the end of the story.

Rebound type two: saving leads to spending

The second type of rebound effect stems from the fact that when we *do* succeed in reducing the energy use of an activity by switching to a more efficient technology, we'll usually end up saving money, even if we also consume a bit more. Some green campaigners have focused strongly on exactly this point: that by increasing energy efficiency we can boost our household budgets or company bottom lines. That's true, of course, but every penny or cent saved on energy becomes available for spending on something else – and that spending also has an energy impact.

The same principle applies to other types of green-minded behaviour change such as turning down the thermostat, or cycling to work rather than driving. Technically, these are 'energy conservation' efforts rather than 'energy efficiency' improvements, as they involve consciously forgoing energy-intensive services rather than trying to get the same services from less fuel. But the net effect is the same: you use less energy and end up with more cash in your pocket for spending on other things. Even if the money saved gets left in a bank it will trigger impacts elsewhere as it will allow the bank to increase its loans to other companies or consumers.

In either situation, the carbon impact will depend on what gets purchased. If the money gets spent on buying a solar panel or enlarging a tropical reforestation project, then the environmental benefits of saving the energy will be multiplied. If, on the other hand, the savings tip the balance of a decision in favour of taking a weekend flight, then the emissions from that trip may be substantially more than the carbon saved in the first place. An individual case could therefore go either way, but *why* the person or company saved the energy probably has a big influence on the nature of the substitute spending. If they saved the energy as part of a conscious effort to be green, they'll be far more likely to use the savings in a low-carbon way. If they did it purely for financial gain they might spend the savings on something energy-intense. We'll come back to this important but overlooked point later in the book.

Individual cases aside, if we look across a whole country at millions of people saving money through efficiency savings and greener behaviours, the money saved will most likely get spent on a fairly typical range of goods and services. A few people may buy extra flights, and a few may buy extra solar panels, but most will spend a bit more on food, or leisure, or gadgets, or whatever else. In this scenario, there *will* usually be an energy saving, because a dollar or pound spent on electricity or driving requires more energy and releases more carbon than the average dollar or pound spent in the wider economy. Nevertheless, the fact that we spend the money elsewhere is – to some degree – almost always another source of rebound.

Rebound type three: others absorb the slack

The third kind of rebound effect lies in the fact that, when one person or company consumes less energy, that will free up more fossil fuel for others to use. Imagine that a whole town of green-minded people all reduce their energy consumption, cutting the amount of natural gas burned (and carbon emitted) at their local power plant. That would be a huge achievement worth celebrating for all kinds of reasons, but it would also put downward pressure on the price of gas and electricity, enabling other homes and businesses to use more. If *everyone* in the country used less power, then sooner or later the country would import less gas, or export more, and national emissions might fall. But as we'll see later, the same wrinkle applies globally: less fuel use in one country enables more use in others.

As with the first two types of rebound effect, it's impossible to know exactly what proportion of our energy savings is cancelled out in this way. Economists can make predictions but they can never be sure of the extent to which an increase in energy use in one place was enabled by a reduction in energy use elsewhere. The dynamics of supply and demand are too complex to pin down, especially given that much of the world's fossil fuel gets used not in cars and power stations but by companies working hard to grow their markets for all sorts of other goods.

One way to think about all this is to imagine the extraction and combustion of fossil fuels – and the carbon footprint of the world's consumers – as being coupled together like the carriages of a train. You can't slow one without also slowing the others, and the three push and pull each other along.

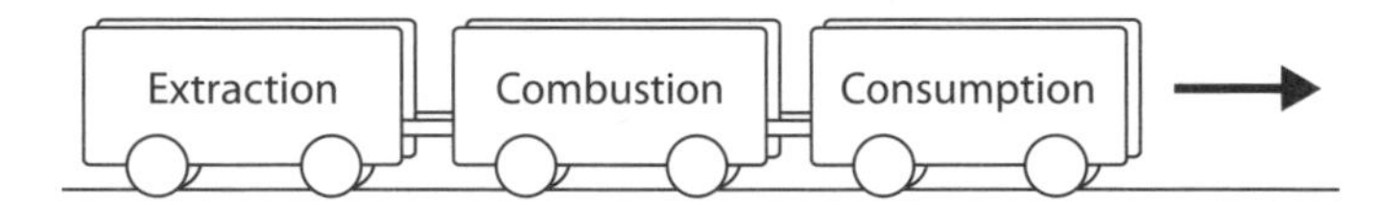

Try to slow the consumption carriage and, even if you persuade consumers to change their habits, you're up against the push of the other two. Companies may mop up the energy saved and create different goods and services instead, doing whatever they can to encourage consumers to develop new 'needs' along the way. Moreover, many people still live in relative or absolute poverty and are understandably focused more on raising their own standard of living than cutting their carbon, so companies can shift *where* they sell as well as what they sell.

But there's a pull as well as a push. Try to slow the extraction carriage by, for example, taxing fossil fuel production and, even if you overcome the fierce lobbying efforts of the companies and countries that produce the fuels, you'll face resistance from the world's energy-consuming households and companies as fuel prices soar. This will undermine political confidence and provide the extractors with the perfect excuse for resisting regulation: they're only producing fuels to satisfy consumer demand.

To have the best chance of cutting emissions, it's clear that efforts are needed to apply the brakes to all three carriages at once. If we focus on just one, the momentum of the other two will make a solution less socially, economically and politically feasible. What is too often missed is the need to put downward pressure on supply as well as demand.

How the Canadian government website justifies its tar sands operation: meeting consumer demand. But supply drives demand as much as demand drives supply.

Rebound type four: social and technical ripples

The three types of rebound we've looked at so far are significant, but they may be nothing compared to the myriad of more complex and indirect ways in which the global economy is affected by energy efficiency gains. We looked earlier at how efficient cars can encourage more driving. A further consequence is that if people drive more, the need to build and maintain highway infrastructure also goes up. And as driving longer distances becomes more affordable it becomes easier for people to live in suburban, semi-rural or other low-density areas. This in turn may lead to bigger houses that take more energy to build, furnish, heat and supply with services such as rubbish collection. The extra houses will also require new out-of-town shops and other infrastructure.[5]

Similarly, swapping a flight for a video call brings a clear and immediate carbon saving. But if the result of inexpensive video conferencing is that more international conversations are had, enabling new relationships to thrive between people living thousands of miles apart, there are bound to be instances of people finding that they want to meet their new-found friends or colleagues face to face. It's perfectly possible, therefore, that video conferencing increases the total demand for flying.

Or take engineering improvements. A new shipping or drilling technology designed to help plant off-shore wind turbines on the seabed might also turn out to be useful for opening up additional deep-sea oil reserves.

A rebound example

To see how all the various types of rebound effects fit together, consider a simple example, illustrated in the diagram below. Assume that a whole country suddenly has access to more efficient cars that double the mileage per gallon. At first glance, that should halve the amount of fuel used. But the new car makes driving cheaper, which encourages or enables some people to drive more than they otherwise would have done (rebound type 1). Overall, fuel use in cars would still fall, but the money saved on fuel will be spent on other things, or put into a bank, where it will get lent to someone else or invested in a company. Either way, the things purchased will usually require energy to produce (rebound type 2).

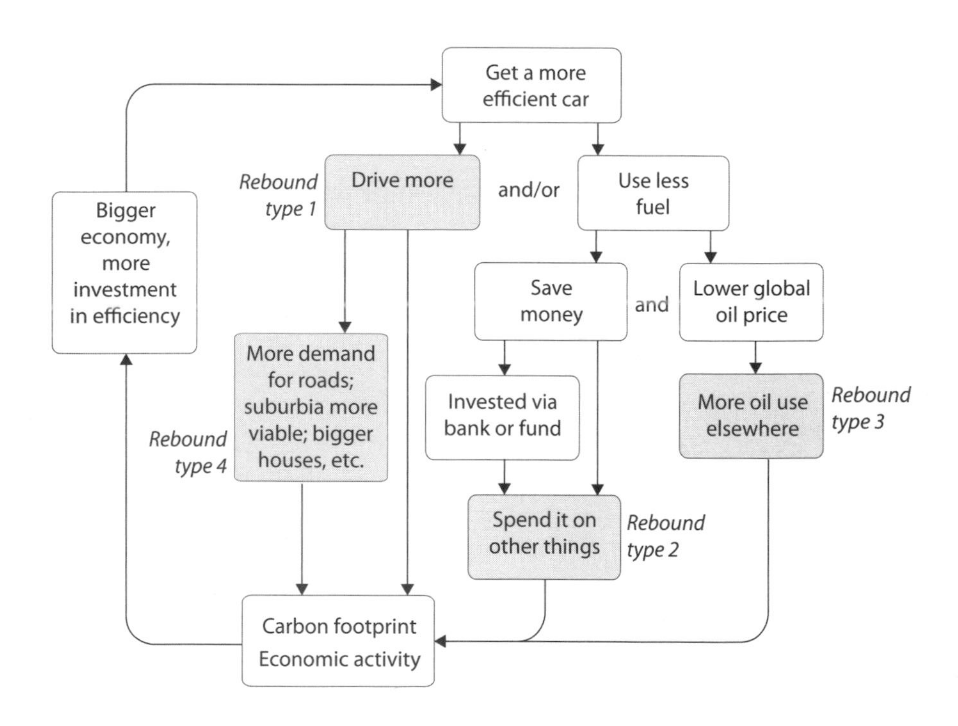

In addition, by saving fuel, the country's drivers will reduce demand for oil, which will tend to lower the price, enabling other countries and sectors to use more (rebound type 3). All the likely paths help drive the global economy forward, leading to more technological development, and perhaps one of the technologies employed in the efficient car engine could be repurposed in a coal mining machine, helping make fossil fuels cheaper to extract. Not just that, but the extra driving may cause other cultural ripples, such as adding to demand for road building and energy-hungry suburbia and also bringing forward the date when the car will need to be replaced (rebound type 4).

It's impossible to put meaningful numbers on how much each of these steps reduces the expected benefits. It might be feasible to estimate the first type of rebound, but when you look at all of them together – and consider the cocktail effect – you're quickly into the realms of guesswork. Overall, though, this example shows how increasing energy efficiency can quite naturally go hand in hand with rising energy use.

Rebounds between countries

Despite all the rebound effects discussed above, the *total* amount of energy used and carbon emitted by most rich countries is falling. Higher global fuel prices and carbon regulation are two of the factors driving this. The EU met its Kyoto Protocol target of reducing emissions by 8 per cent in 2012 relative to 1990 levels. Emissions in some countries, such as the UK, have been falling for decades, and in recent years the downward trend has been observed in the US. According to the latest figures from the IEA, America's emissions from energy use fell by 1.7 per cent in 2011, in large part thanks to a switch away from coal towards shale gas and renewables. The EU figure was similar and OECD nations as a whole made cuts of 0.6 per cent. These savings are very encouraging but so far the global carbon curve has remained on its unflinching exponential path. In other words, gains in the rich world have been almost exactly counteracted by increases in the developing world. That's not necessarily surprising given that we live in an era of globalised markets, with huge volumes of both fuels and goods traded between nations, enabling rebound effects to work across national borders.

Looking first at flows of fossil fuels, it's clear that reductions in oil, gas or coal use in the rich world are related to increases in emerging economies. The causality runs both ways: as developing countries have grown, this has increased demand for fossil fuels, pushing up prices and encouraging the rich world to use less. In parallel, rich countries have started actively to try to reduce their emissions, freeing up more fuels for use in emerging economies. In some cases, the global links are more obvious as governments seek to reduce their own emissions at the same time as maximising their exports of fossil fuels for use elsewhere. In 2012, Australia introduced a carbon tax but also began debating plans to build a series of mega-mines to export more coal overseas.[6] Similarly, as new coal power stations finally became uneconomic in the US due to other energy supplies coming on stream, American coal exports increased and firms started lobbying for new west-coast ports to help ship even more of their fuel to China.[7]

Even the UK, with its world-leading target of cutting carbon by 80 per cent by 2050, has a stated aim of 'maximising the economic recovery of oil and gas from the UK's oil and gas reserves'.[8] As we were writing this chapter, the country's self-styled 'greenest government ever' announced a new pot of public money to subsidise technologies for boosting gas and oil extraction in the North Sea.[9] In the same month, on the other side of the Atlantic, the two participants in the US presidential election sparred over who would do more to increase domestic oil and gas production. As journalist George Monbiot put it, policymakers are simultaneously seeking to 'minimise the demand for fossil fuels and maximise the supply'.[10]

Fuels aren't the only thing that flow between countries. We also have global markets in commodities, manufactured goods and services. Whereas people in Germany may once have bought toys or clothes made in European factories, these goods are now more likely to come from China's Guangdong Province. British consumers talk to tech support staff in Bangalore, and Americans rely on Mexican *maquiladoras* for manufacturing. Since all these goods and services take lots of energy to provide, the result of this shift is rising emissions in the developing world. But whose emissions are they? At the moment, the global climate talks work on the assumption that the emissions are the responsibility of the exporting nation since the carbon is released within its borders. But it's obvious that much – some would say all – of the responsibility for this carbon ultimately lies with the nation that imports and consumes the goods.

To get a more meaningful picture of global emissions, a team of academics crunched a huge amount of trade data to work out the total carbon footprint of consumption in each major country, including imports and excluding exports. (In the jargon, this is called 'consumption-based emissions reporting', as opposed to conventional 'production-based reporting'.) What they found was remarkable. Between 1990 and 2008, developed countries as a whole cut their carbon emissions by 2 per cent, but their total carbon footprint in the same period actually *grew* by around 7 per cent.[11] As expected, the rise was higher than average in the US, where climate policies have consistently met with strong political opposition, but even in climate-progressive Europe, the apparent 6 per cent cut in carbon emissions was slashed to become just 1 per cent. The situation is even starker in the UK, which in the past two decades has become unusually reliant on imported goods. Emissions within Britain's borders fell by 19 per cent between 1990 and 2008, giving, on the face of it, a reassuring story of successful carbon reduction in line with national targets.[12] But Britain's total carbon footprint rose by around 20 per cent in the same period once imports and exports were factored into the equation. A major saving turned out to be a substantial increase.

It's worth noting that the academics who did such a heroic job of estimating the footprint of the world's imports and exports were only able to do so by making lots of assumptions about how goods get produced in each country. In reality, the carbon footprint of China's exports may well be even greater than the academics assumed due to the carbon intensity and inefficiencies of the country's industries.[13] Furthermore, a recent study showed that Chinese emissions in recent years have been a remarkable fifth higher than the official statistics show.[14] For both these reasons, imported goods are probably an even bigger factor than the published research suggests.

The shift of fuels and manufacturing from developed to developing countries hasn't generally been the perverse result of environment policies. Though some industries may have relocated from, say, Europe to China due to green taxes, the exodus of heavy industry from mature to emerging economies also reflects a host of broader factors such as labour costs. Hence the trade in goods and services isn't necessarily a 'rebound effect' in the conventional sense. There's no doubt, however, that imports and exports have blurred and undermined the developed world's progress on reducing emissions. And that's only the start of it. Importing goods from emerging economies has also been absolutely key to boosting economic growth *within* the exporter nations, helping drive their domestic markets for goods and services too. That's good for economic growth and poverty reduction, but from the perspective of carbon emissions it's yet another example of a ripple effect that has helped sustain the global carbon curve's exponential path.

One ray of good news from the UK is that its government is starting to recognise this issue. A recent House of Commons Select Committee report concluded that climate policymakers should consider the total carbon footprint of a nation's consumption rather than only looking at its direct emissions.[15] If the rest of the world follows suit and adopts consumption-based emissions reporting alongside the traditional production-based approach then traded emissions will be visible to all. That would help clarify national and regional progress – or lack thereof – and make a global climate deal a fraction easier to agree upon.[16] It would also open up potentially important new policy options. For example, if nations applied their domestic carbon regulation to imported commodities or other goods, that could avoid the balloon-squeezing effect *and* create an incentive for exporter nations to cut their own emissions. As we'll see later, this would open a political and practical can of worms, but given that the future of the planet is at stake, such policies are at the very least worth seriously considering.[17]

Rebounds and the curve

With so many ripples and rebounds at work, trying to quantify the overall global impact of any efficiency gain or local carbon saving is impossible. The effects are too numerous, too complex and too subtle. But what is clear is that individuals, companies, countries and whole continents are intricately interconnected. What one of us does will impact on what the others do and very often those impacts will offset some or even all of the carbon savings. Viewed this way – and in the context of the unchanging global carbon curve – it looks like rebound effects are extremely important. So much so that efficiency improvements and piecemeal savings can't be relied upon to ever solve the problem in themselves. (As we'll see in the next chapter, something similar applies to clean energy systems as well.)

Only if emissions are explicitly constrained – for example, through a global carbon cap-and-trade scheme – can we be sure of staying within any given carbon budget. If carbon was capped globally, then the whole problem of rebounds would disappear. We'd know for sure that the rebound effect of any energy-saving action or technology was precisely 100 per cent but that wouldn't matter. Indeed, with a global cap, energy efficiency improvements would serve a different role. Rather than trying to reduce global energy use, which they've consistently failed to do over the decades, they'd be about maximising the benefit that society could extract from each unit of energy available. Similarly, voluntary energy savings wouldn't be about reducing total fossil fuel use; instead they'd be about minimising waste and enabling those with greater need, such as those in poor countries, to burn the available fuel instead.

Although few people realise it, this is already true to a degree in regions where there's a local carbon cap, such as Europe. Across the EU, emissions of power plants, steel mills and other big industries are capped as part of the European Trading Scheme (ETS). Because the total carbon output is fixed in advance, it doesn't necessarily cut any carbon when we save electricity in our homes and offices. Instead, any permits saved by the local power station can be sold for use in a car factory or steel plant elsewhere in the continent and total emissions remain unchanged.[18] But even if the ETS was well designed and ambitious (which it isn't), a regional carbon cap wouldn't deal with the carbon flows between nations in the form of fuels and goods. To work, we need the cap to apply globally.

The real role for efficiency, therefore, is as an enabler for tightening up global regulation on fossil fuel use. By driving up efficiency *in parallel* with clamping down on carbon emissions, governments can make the whole process as inexpensive and painless as possible. That, combined with efforts to reduce the emissions from exported fuels and imported goods, can help make a global deal more likely, as we'll see later. But due to rebounds, we shouldn't imagine – or allow policymakers to imagine – that efficiency in itself will cut emissions. There's even an argument to say that in the absence of carbon regulation rising efficiency may be counterproductive, as it squanders an opportunity to inexpensively cut carbon and instead feeds the global energy feedback loop.

As for voluntary carbon and energy savings at the personal, corporate and national level, their real value may be less about reducing emissions directly and more about showing leadership and changing cultural norms in a way that would make a global deal more likely. Any family or firm or nation which takes steps to save energy sends a message to the wider world that climate change is important and that solving it is manageable. But if we accept that our green efforts are as much about creating social and political ripples as they are about directly stemming the flow of carbon into the atmosphere, that suggests we may need to think more about maximising those ripples. Companies that genuinely want to help tackle climate change, for example, need to focus not just on measuring and fine-tuning their carbon footprint but also on examining the signals they're sending to customers and politicians through marketing and lobbying. Similarly, individuals can to make demands at the political level as well as trying to reduce energy use directly.

7. People, money and technology

How balloon-squeezing also affects population, economic change and alternative energy

Efficient technologies and voluntary carbon savings are one way of seeking to reduce demand for fossil fuels. Other proposed solutions include installing more renewable or nuclear energy to displace oil, coal and gas use, and slowing the growth in population or economic activity to help restrain energy use overall.

Many academics and policy wonks have pointed out that all these approaches fit together, arguing that climate change is caused as a result of *people* spending *money* in ways that trigger the *energy use* that causes *emissions*. This can be expressed by a simple equation. Emissions from energy use = the number of people × the average GPD per person × the average energy used to create each unit of GDP × the average emissions caused by each unit of energy. Or, to put it more snappily:

$$\text{Emissions} = \text{population} \times \text{affluence} \times \text{energy intensity} \times \text{carbon intensity}$$

This equation, sometimes known as the Kaya Identity, can provide insights into what has been driving the carbon curve so far, and how much we might need to reverse or accelerate particular historical trends.[1] The nice thing about the Kaya approach is that the equation is mathematically watertight: whatever humans do or don't do to solve climate change, those four components will always multiply together to give total emissions from energy use. But that doesn't mean this way of thinking can't be misleading. All too often, experts working with this equation assume that the four elements don't really interact, so that, for example, a 5 per cent drop in the carbon intensity of energy will lead to a 5 per cent drop in global carbon output. The problem is that, as we'll see, changing one component can change the others in the opposite direction. Kaya therefore gives us another way of looking at how balloon-squeezing effects cancel out apparent progress.

This chapter explores the four components in turn to see how each has helped drive emissions so far and how much each one could help solve the problem going forward. Is population the key to solving climate change, as some people claim? Or economic growth? Or clean energy?

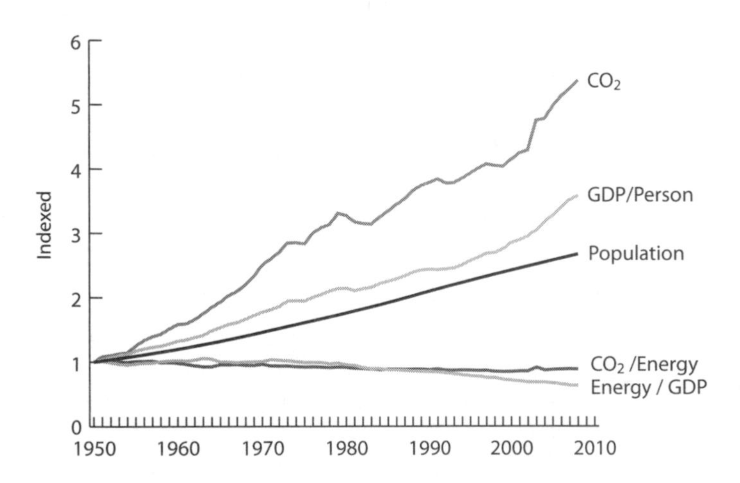

The four Kaya components tracked since 1950.[2] These multiply together to give total carbon dioxide emissions from energy use, which is also shown on the graph. For carbon emissions to fall in absolute terms, any increases in population and affluence need to be outpaced by reductions in energy intensity and carbon intensity.

Population

No issue in the climate change arena is more divisive and controversial than population. Read the comment thread under any environment news story, or listen to the questions at the end of any climate panel discussion, and it's clear that in some people's minds, this is the elephant in the room. There are simply too many of us, and only by levelling off and then reducing the population can we solve global warming.

We showed earlier how growth in population and energy use have gone hand in hand over the centuries – and as we'll argue later, it's almost certainly true that solving climate change would be easier with fewer people. But a quick glance at the data from the last few decades suggests that focusing on population is no panacea to reducing fossil fuel use.

Reliable estimates of global population stretch back to 1750. They tell us that it has never had truly exponential characteristics. For a long time the trend was 'superexponential': the annual *rate* of growth rose steadily throughout the nineteenth and the first two-thirds of the twentieth century, quickly bouncing back from blips such as world wars. But all that changed abruptly in the late-1960s. For the last few decades the annual percentage growth rate has been dropping like a stone and the number of additional people added to the world each year has been just about constant. This reflects an unprecedented reduction in global fertility levels. In the 1960s, the average woman had five children. Just over fifty years later – in what must count as one of the most profound societal changes in human history – that figure has fallen in half and is expected to fall further over the coming decades.[3]

Interestingly, however, this slowdown in population growth hasn't helped us deviate from the exponential emissions curve. That suggests the other Kaya components must have reacted in a way that compensated for changes in the population trend.

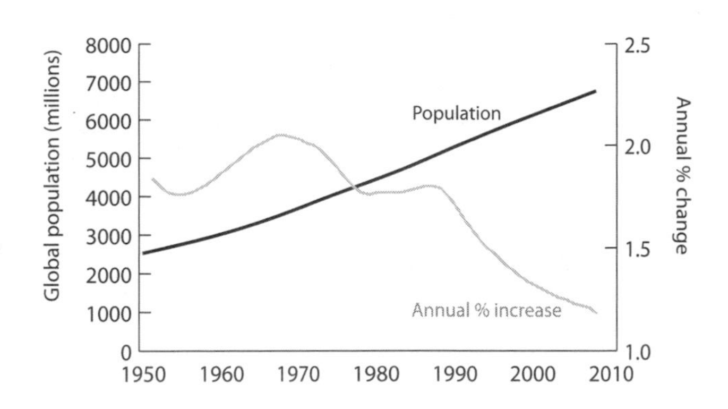

On closer consideration, that is perhaps not surprising. For one thing, population growth and income growth are interrelated. In the so-called 'demographic transition', which is a keystone of population studies, each country starts with a high birth rate and a high death rate, giving a stable but small population. As sanitation and healthcare and agriculture improve, death rates fall, driving up the number of people. Eventually education, urbanisation and contraception kick in, incomes rise, the birth rate falls too and population levels off. Since rising incomes and rising carbon footprints tend to go hand in hand, it makes sense that a slowdown in population growth may get cancelled out by an increase in consumption per head.

Another (related) reason why future changes in population growth might not have the impact one would expect is that birth rates vary so much around the world – and tend to be highest in the countries where carbon footprints are low. The number of people in Sub-Saharan Africa is growing at around 2.5 per cent a year, a full three times faster than in developed nations.[4] But the average Sub-Saharan African lifestyle takes less than a tonne of carbon dioxide a year to provide. That compares to around six tonnes globally, fifteen tonnes in much of Europe and more than twenty in North America and Australia.[5] The differences in wealth and footprint per person are so great that even if the whole of Sub-Saharan Africa disappeared in a puff of smoke tomorrow, emissions would dip by only a few per cent – and perhaps even less if consumers in other countries picked up the slack in the fossil fuel markets.[6] As a 2009 report by the International Institute for Environment and Development put it:

> A significant proportion of the world's ... population have consumption levels that are so low that they contribute little or [no] emissions. Reviewing carbon dioxide emission levels for nations and how they changed between 1980 and 2005 ... there has been little association between nations with rapid population growth and nations with rapid greenhouse gas emission growth; indeed, it is mostly nations with very low emissions per person (and often only slowly growing emissions) that have had the highest population growth rates.[7]

None of this is to say that population is irrelevant to solving climate change. Common sense backed up by a range of academic studies says that more people will lead to higher levels of future energy demand.[8] If the world could agree a global carbon cap, that might not matter, as the extra demand could only be met by renewables or nuclear; but without one, all that extra demand will doubtless push up fossil fuel use yet further as livelihoods improve in the poorest countries. Not just that, but more people will tend to mean more emissions from deforestation and farming, as we'll see in chapter twelve. In addition, a rising population will make global warming more dangerous for the simple reason that most of the additional people will be born in areas that are unusually at risk from extreme weather events and dwindling agricultural yields, and which are least economically equipped to deal with the impacts.

There's plenty to play for here. The UN's central population forecast for 2050 is a little over nine billion, but its high and low scenarios range from around eight to ten billion. By the end of the century, the main projection rises to ten billion and the possible range widens even further, stretching from just over six billion at the lowest end to nearly sixteen billion at the highest end.[9] Whether these predictions are realistic is a matter of debate: author and journalist Fred Pearce has reported that the UN may have artificially inflated its central projection to avoid making the world complacent about population growth.[10] But it *is* clear that what happens to fertility rates over the next few decades could make a very significant difference to the timing and size of the world's population peak. As the UN puts it, 'small variations in fertility can produce major differences in the size of populations over the long run'.[11]

For all these reasons, sensible policies to restrain population growth – such as much greater efforts to help women around the world avoid unwanted pregnancies and get access to education and the workplace – are no-brainers. Nonetheless, the evidence so far suggests that limiting or even reversing population growth may not in itself reduce carbon emissions as much as we might expect. There would be fewer extra people but not necessarily fewer fossil fuels. The balloon would change shape but not necessarily shrink.

Affluence

Affluence in this context means GDP per person, which roughly equates to the average person's spending power in a typical year.[12] That includes spending on all kinds of things: food to furniture, healthcare to haircuts, cars to computers. Despite the continued gulf in incomes between the world's richest and poorest, average global affluence has been rising fast (and

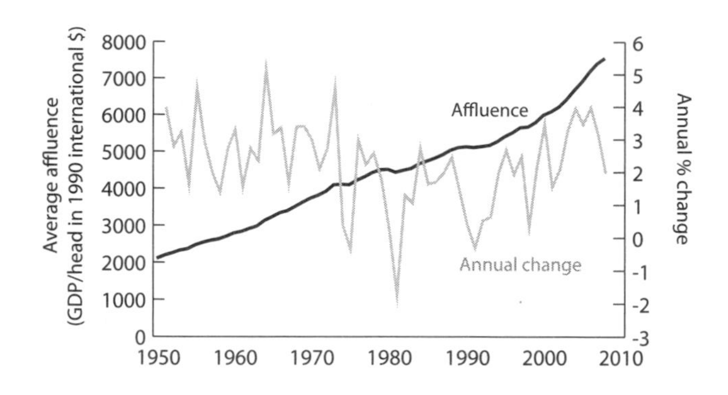

reasonably steadily) in recent decades, by an average of 2.2 per cent per year.[13]

Lots of different factors have contributed to this growth in economic activity per person – from expanding global markets and greater specialisation of workers and nations to more productive technologies. Different economists and thinkers have different views on the relative significance of each driver, but as we argued in chapter one, it's clear that the availability of energy has been closely bound up with it all.[14] The Kaya components appear to confirm this: since 1950, and plausibly since the beginning of time, affluence and carbon emissions have moved in lockstep. When affluence has risen, so have carbon emissions. When (and only when) affluence has fallen – such as in the US during the 1970s oil crises, the Soviet Bloc after the collapse of communism, or the world after the financial crash in 2008 – so have emissions also declined.

That affluence and carbon emissions should be closely bound up is no surprise because almost all goods and services take energy to provide. Energy powers the machines that have driven up labour productivity, the trucks and ships that have allowed global markets to expand, and the mining operations that provide all the raw materials. As a result of this, some commentators and many in the green movement have argued that slowing, stopping or even reversing GDP growth may be the key to reducing emissions and other environmental impacts – or an inevitable consequence of doing so. A version of this discussion has been running at least since the 1970s when economist Herman Daly published his book *Steady-State Economics*. But the debate has been particularly lively in the climate change sphere since Tim Jackson, a British academic economist, called for rich countries to abandon the quest for ever-rising GDP in his influential 2009 book *Prosperity Without Growth*.

As we'll explore in the next chapter, it's certainly possible that solving climate change by leaving most of the world's fossil fuel reserves in the ground could have an impact on economic growth. But *aiming* for zero growth in rich countries as a means of solving climate change doesn't look hugely promising. For one thing, abandoning growth in rich countries alone probably wouldn't cut carbon by a huge amount – at least not directly. What we traditionally call the 'developed world' – the US, Canada, Europe, Japan, Australia and New Zealand – now accounts for a minority of global emissions. And these countries would still produce plenty of carbon dioxide even if their economies weren't growing. Indeed, one recent paper showed that emissions rise faster in economic boom times than they fall during recessions, the most plausible reason being that halting growth does little to change the way we use our existing cars, buildings and power stations.[15]

Even if economic growth were halted permanently in rich countries, the carbon savings might be smaller than expected if the squeezed balloon bulged in the developing world. It's even possible for the reasons described earlier that a slowdown in global affluence would lead to faster population growth, offsetting yet more of the gains.[16]

There's also a more fundamental challenge with aiming to slow or reduce GDP as a means of cutting carbon. Even if policymakers had the will, it wouldn't necessarily be within their powers. Economic growth is driven at least in part by innovations in technology and processes that drive up productivity by increasing the amount of goods and services produced by each worker, acre, building, or tonne of raw materials. Container shipping, production line factories and the internet are just a few prominent examples. This implies that for a zero-growth economy to exist, the march of technology would need to be slowed or stopped. It's not immediately clear how a government could do that (at least not without radical changes to things such as property laws) and neither would it obviously be a good idea, given that a technological revolution is needed in the world's energy systems.

Even if none of this were true and governments *were* able to level off growth deliberately, they'd also face a serious political challenge. Historically, when the economy has shrunk, debtors have defaulted, investors have lost their money, lending has dried up and a recessionary spiral has led to unemployment, failing tax income, cuts, pain and social unrest. The implication is that modern economies are only stable when they're growing. There may be some things governments can do to limit these effects – such as reducing the working week to share out the available employment or aiming for lower levels of consumer, corporate and national debt – but it's not clear these would be enough to make a zero-growth economy politically feasible.

For all these reasons, seeking to limit affluence doesn't look like an easy or reliable way to solve climate change. But critics of economic growth *do* have an important point to make. As we'll discuss later, one of the key reasons why policymakers aren't doing more on climate change – including at the global level – is they're terrified that constraining fossil fuels will slow or reverse the growth in GDP. So whatever else we do, to solve the problem we'll certainly need to stop *prioritising* short-term economic growth above all other considerations. But that alone won't solve the problem.

Energy intensity of the economy

The energy consumed to create each unit of economic activity has been falling for decades. In fact the rate of improvement has itself been improving, reaching an annual reduction of 1.5–2 per cent per year since the turn of the twenty-first century. In other words, we're becoming ever more energy-efficient at generating economic activity. This is partly the result of mechanical and technical improvements, such as LED bulbs and hybrid cars, but the key factor is almost certainly that an ever-larger proportion of our economy revolves around services and knowledge rather than physical products. We are spending proportionally more on 'virtual' things such as digital media, telecommunications, education and personal services of every kind; and proportionally less on energy-intensive items such as fuel, electricity, cars and building materials. Despite problems in the fifties and sixties, it looks likely that a similar if less clear and consistent trend extends much further back.[17]

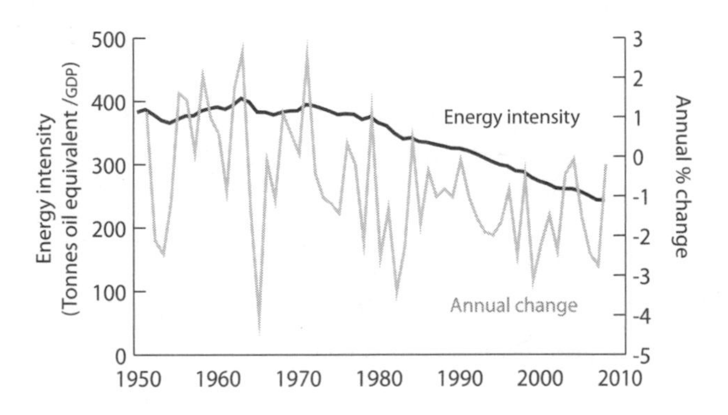

However, before jumping to the conclusion that the transition to a knowledge- and service-based economy can help us mitigate climate change, there is a key point to emphasise: a *relative* shift towards a more virtual economy does not necessarily lead to an absolute fall in energy use. In other words, it is not at all clear that the ever-rising level of economic activity based on knowledge and services stands to replace rather than be in addition to, and reliant upon, an ever-expanding base of heavy industry and physical products. Indeed, as the Kaya graph at the start of this chapter shows, energy intensity is the only one of the four components that is making any serious headway in the downward direction, with population and affluence growing more than fast enough to compensate. The result is that – although emissions are growing slightly *slower* than total GDP – their exponential increase remains unchanged. This suggests strongly that any virtualisation that the global economy has undergone so far has involved adding services, not replacing the industrial underbelly.

Looking forward, there's huge scope for reducing energy intensity further by cranking up efficiency standards and finding technologies and economic incentives that can help change behaviours to reduce energy waste. Some models of future energy trends suggest that these kinds of improvements could allow us to get our goods, services, heating and power from as little as *half* as much energy by 2050. But as we saw in the previous chapter, in the absence of a carbon cap, such measures would be subject to rebound effects and could even enable us to consume yet more energy at the global level.

Carbon intensity of energy

The three components explored so far multiply together to give total global energy use, leaving just the carbon intensity of that energy to consider. If we were moving fast enough towards some mix of renewable and nuclear power, or fossil fuels burned with carbon capture systems, and if we were sure these sources were replacing rather than supplementing unabated oil, coal and gas use, then emissions would be falling and energy consumption would cease to be a problem. So what's going wrong?

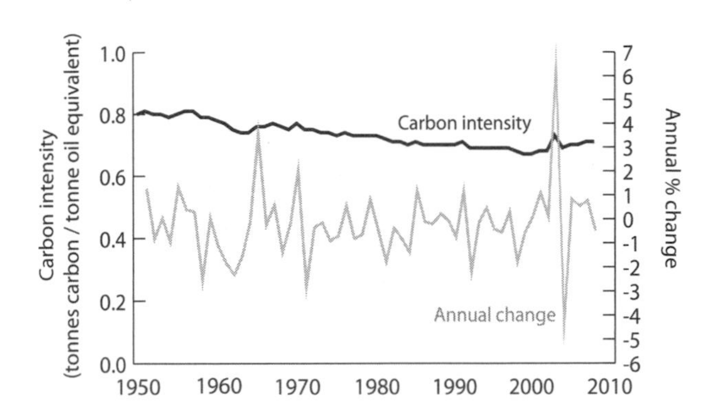

The news here is a little surprising – and not in a good way. For all the headlines we read about alternative energy sources, the carbon intensity trend over the last few decades has been just about flat. As the graph shows, it fell slightly through the 1990s but since the turn of the century has been fairly level or possibly even creeping back up, despite all the efforts to promote renewables. As a result, the carbon emissions produced from the average unit of energy are no lower today than they were thirty years ago – and not much lower than in 1950.

At first glance, this seem odd. We've all seen a gradual increase in the number of wind turbines and solar panels and heard talk of investment in other energy sources such as marine power and new nuclear plants. But for now the total increase in low-carbon power is small beer compared to the growth in fossil fuel use, including a very significant shift towards coal, by far the dirtiest and most carbon-intensive major energy source.[18] The chart overleaf shows the new energy capacity that came on-stream globally between 2001 and 2011. Coal dwarfs the other sources, and fossil fuels together outstrip renewables by almost an order of magnitude. The additional coal was burned almost entirely in Asia – especially China – though much of it was mined elsewhere and a good deal of it used to produce goods for the rest of the world.

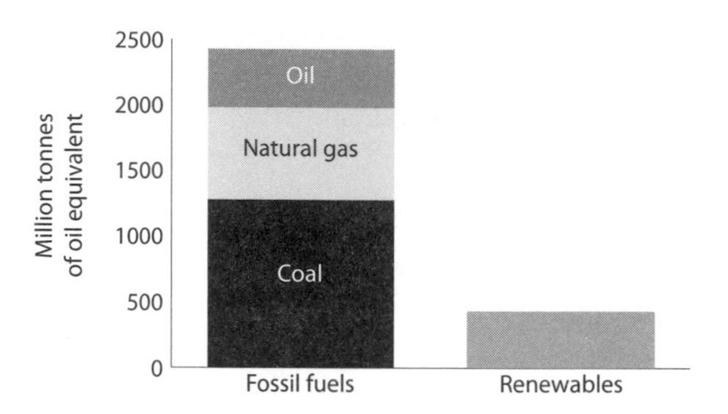

Net *additions* to the world's energy systems in the period 2001–2011.[19] Fossil fuels account for around four-fifths of the total. Nuclear isn't shown because its total output was lower in 2011 than it was in 2001.

Looking at *rates* of growth in different energy sources reveals a more positive story. Although low-carbon energy is still being outpaced in absolute terms by fossil fuels, it is growing much faster. In fact, solar and wind power are growing not just fast, not even just exponentially, but superexponentially: the rate of increase is escalating. In the early 1990s, renewables (excluding hydro) grew around 5 per cent per year. By the turn of the millennium, the figure was around 7 per cent. By 2010, it had increased to 17 per cent.[20] Such a rocket-powered trajectory probably isn't sustainable for very long, but the direction of travel does suggest that low-carbon energy will soon represent a much more significant part of the global energy picture. In parallel, prices for clean energy sources are coming down, while fossil fuel prices, on average, are going up.

This is encouraging but once again we need to be cautious of balloon-squeezing effects. Policymakers tend to assume that a unit of low-carbon energy will automatically obviate the need for a unit of fossil fuel capacity, but that's only true if energy demand is unaffected by energy supply. In reality, as we saw in chapter one, global energy consumption has been rising

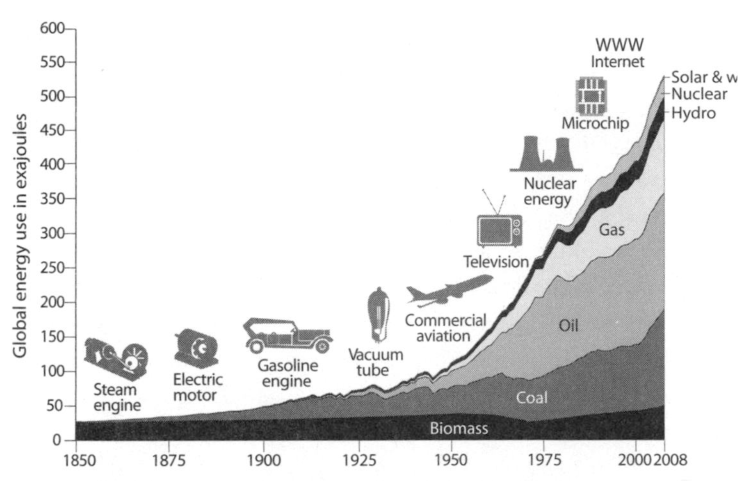

Total human energy consumption since 1850. The chart shows that, in themselves, new energy sources don't tend to reduce use of the existing sources. Instead, they drive economic activity and increase global energy use.[21]

exponentially for hundreds of years and, as the graph above makes abundantly clear, as each new source of energy has come on stream, we've continued to use the existing ones in growing quantities. Renewables and nuclear are the same: there's no real evidence yet to show that they have been *replacing* fossil fuels at the global level.

At the national level it's possible to square the circle by building clean energy capacity *and* constraining fossil fuel use with taxes or carbon caps. But even that won't ensure that the oil, coal and gas stay in the ground, as the owners of the reserves may simply extract the fuels and sell them in another country. Without a global climate deal or national taxes on fossil fuel production or exports, this is surely inevitable, at least to a degree, because globally there is no shortage of energy demand. Just ask any of the billions of people living and working in non-air-conditioned buildings in the tropics, or the majority of people who have never taken a holiday flight.[22] As people continue to strive for greater levels of material prosperity, the capacity of the developing world to mop up fuels not burned in richer nations may be almost limitless in the absence of any carbon constraints – at least for the next decade or two.

There may even be situations where it's possible to trace a link between more low-carbon energy and *more* fossil fuel use. A case in point is small-sized next-generation 'fast' nuclear power plants. On the one hand, these could in theory be enormously useful for solving climate change, producing in a best-case scenario vast quantities of low-cost low-carbon power and even consuming existing plutonium waste as fuel. But according to one leading exponent of the technology, the earliest adopters of the reactors are likely to be the companies that run giant natural gas pipelines, such as the ones that take Russia's gas to Europe. At the moment, a slice of the gas transported has to be siphoned off along the way to power the turbines which drive the remaining gas down the pipeline. If these could be replaced by modular nuclear reactors then such waste could be avoided, increasing the commercial viability of ongoing gas supplies for the world's boilers and power plants.[23] This example may be an anomaly but more broadly it's possible that adding more energy of any type will boost the world economy, driving the feedback loop explored in chapter one and thereby increasing demand for and access to all other forms of energy, including fossil fuels.

Advocates of low-carbon energy point out that renewables and nuclear are already as cheap as fossil fuels in some situations and will eventually undercut them globally. Unfortunately, it doesn't look like there's much chance of this happening anywhere near fast enough to avoid unacceptable risks from climate change – at least not without almost unimaginable technological breakthroughs. Anyhow, even when the costs become comparable there is no guarantee we'll stop burning the oil, coal and gas. For one thing, figures comparing the costs of energy sources can be misleading because it isn't the *average* cost of a unit of energy that determines which source gets used but the *marginal* cost – the cost of creating one more unit in any given situation.[24] If a company already owns a coal mine, a coal train and a coal power plant, then the marginal cost of continuing to use them is going to be low, even if the total cost of coal power tends to be higher than total cost of wind. Similarly, if a household owns an expensive conventional car, it might be cheaper or more practical to keep buying petrol for that car than it would be to replace it with an electric model, even if the cost per mile of driving is higher. For that reason, we clearly need to think not just about the fossil fuel reserves but all the infrastructure and machines we have for burning them. More on this later.

Another challenge with trying simply to undercut oil, coal and gas in an open market is that current fossil fuel prices often include a generous profit margin. A barrel of Saudi oil might sell for $100 in the global markets but cost only a few dollars to produce.[25] So if clean energy started to undercut that oil, there's plenty of scope for the price to come down. To make fossil fuels entirely uneconomic – so that their owners would be better off abandoning rather than extracting their reserves – renewables and nuclear would need to fall to extremely low levels.

The IEA's modelling doesn't take all of these effects into account but even so its scenario for the next twenty-five years based on current national policies sees huge increases in renewables outstripped by even larger gains in oil, coal and gas. The share of fossil fuels in the world's primary energy supply would fall by only a small amount, from 81 per cent in 2010 to 79 per cent in 2020 and 75 per cent in 2035.[26] As a result, the carbon intensity of energy won't fall far either – and emissions will continue to soar by almost a quarter over that period.

To be clear, we are not saying alternative energy sources aren't important. It's obvious that phasing out fossil fuels will only be practically and politically possible in parallel with a monumental effort to scale-up low-carbon energy sources. But in the absence of a major global effort to constrain fossil fuel use, it's not clear that low-carbon energy *in itself* is helpful.

The same argument applies to natural gas, which is increasingly described as a green option. Since gas emits roughly half as much carbon dioxide per unit of energy as oil does and even less compared to coal, many economists and politicians now see 'fuel switching' to gas – including unconventional sources such as shale gas – as an inexpensive means of rapidly cutting carbon. It's true that where gas displaces coal the benefits in terms of emissions (and air pollution) can be large. The obvious example is the US, where a massive shale gas boom has helped reduce emissions by undercutting coal. But as we saw earlier, US coal exports have increased significantly in the period so at the global level, the natural gas has been at least partly, and perhaps mostly, additional. Anyhow, the choice of what gets exported and what gets consumed domestically is partly a matter of historical accident: the US has plenty of coal export facilities but almost no major gas export terminals. (That looks set to change soon.[27])

Under a global carbon cap, natural gas would almost certainly have a role to play as a low-cost transition fuel. This would happen by itself because, with a high global carbon price, coal would quickly become uneconomic. But given the steepness of the carbon cuts needed, gas can only play a short-term interim role. Any new gas infrastructure therefore needs to be put up on the understanding that much of it will need to be retired or adapted with carbon capture well before the end of its working lifetime. More fundamentally, in the absence of a global constraint on carbon, getting more gas out of the ground almost certainly leads to additional emissions by increasing the global energy supply. And that's before you consider any methane that leaks into the atmosphere as a by-product of shale-gas drilling. The significance of these 'fugitive emissions' is debated, but one study estimated they were sufficient to make shale gas as carbon-intensive as coal.[28]

Four components or just one?

It's a shame there isn't good quality data on all four components going back further than 1950 for us to see more about how they have related to each other through history. But we *can* say that during more than a century and a half of exponentially rising carbon emissions each one has taken a different route. Gains in one component therefore appear to have been compensated for by losses in others. Why would this balloon-squeezing effect be happening? The most likely explanation is that global emissions are ultimately determined not by the multiplication of the Kaya factors, as is sometimes assumed, but by some other overarching factor – such as our ability to extract oil, coal and gas and produce the infrastructure to burn it all. In other words, the Kaya components don't *drive* fossil fuel extraction, they fit around it. This chimes with the theory posited earlier that energy begets energy because it suggests that, at a global level, we simply take out of the ground as much fossil fuel as we can. (Or, in the case of the oil, as much as the OPEC nations allow us to.[29]) Once the fuels are in the energy markets, they get burned, so if one of the four components rises or falls, then the others naturally move to accommodate the fuel use elsewhere.

If this is correct, it's a disaster for any forecasting that assumes the four components are independent – and it shows yet again that to tackle climate change we'll need to focus our attention on slowing the flow of fossil fuels into the global economy. That will require global politics, since no profit-seeking company or self-interested nation is going to abandon its fossil fuel reserves voluntarily. Progress on each of the four Kaya components may help make an ambitious deal more possible, however, provided the others are not allowed to bounce up in response. Slowing population growth will mean that the world's carbon budget won't need to be as thinly spread. Deprioritising short-term economic growth would make a deal more economically acceptable. And investing much more in research and development for clean power sources and energy-efficient technology will make it more politically feasible to constrain oil, coal and gas since it will allow cuts in fuel use to go in tandem with alternative ways of achieving the same levels of utility. Once again, bottom-up progress can help unlock top-down action, but it won't be enough. There is no escaping the need to deal with the underlying problem head on: the extraction and burning of fossil fuel.

PART THREE

What's stopping us?

Why we are finding it so hard to leave fossil fuels in the ground

8. The write-off

The fossil fuel reserves and infrastructure that could be devalued by tackling climate change

In part one we saw how much momentum lies behind global fossil fuel use – and how quickly we need to turn the ship around. In part two we saw how the global socio-economic system flexes and responds to nullify or minimise the benefits of many of the actions we assume should help. An ambitious global climate deal could slam the brakes on the world's fossil fuel flows by capping the world's carbon emissions or imposing a tax on fossil fuel production or use (we'll look more at these options later). But efforts to agree such a deal are up against huge resistance, since many businesses, governments and people are more concerned about what they have to lose from any restrictions than they are about climate change itself.

Powerful extractor nations and companies own oil, coal and gas reserves, many of which would need to be written-off or devalued if an ambitious global deal was struck. Industries, governments and households own plants, vehicles, boilers and other devices that need fossil fuels to function. Ordinary consumers maintain or aspire to lifestyles that entail large carbon footprints, whether that means flying regularly or enjoying low-cost goods and services that take energy to provide. All of this – and the stability of the whole global economy – may be at risk if world leaders decide collectively to phase out fossil fuels at the rate required. With so many interests at stake, most policymakers are reluctant to rock the boat, even though they agree that the global society is walking towards a cliff. Elected governments are nervous of getting voted out, and unelected ones fear anything that will loosen their grasp on power. At the moment, in fact, they're not just failing to constrain fossil fuels; in many countries they're actually *subsidising* them.

This chapter and the next take a look in more detail at these various forms of entrenchment and resistance. We'll start with the physical forms – the fossil fuel reserves, power stations, cars, boilers, tractors, factories and companies that might be devalued by addressing climate change – and then move onto the perceived economic, political and social risks that could be holding us back.

The fossil fuel write-off

Green campaigners sometimes describe the world as being addicted to fossil fuels. If that's true, it is striking that efforts to solve this addiction have focused almost entirely on the users of the substances in question, not the dealers that produce them and bring them to market. Indeed, while people and companies have been asked to change their habits and wean themselves off fossil fuels, the kingpins of the supply network – the companies and countries that own all the oil, coal and gas – have so far been largely ignored. That surely needs to change, because the world's fossil fuel companies and their valuable hydrocarbon assets sit at the very core of the problem.

We saw earlier that the world already has more than enough carbon in its proven fossil fuel reserves to push the climate over a cliff. But many companies – often pushed hard by institutional investors concerned about 'reserve replacement ratios' – are spending billions prospecting for more fuel and building capacity to increase their rate of extraction. One recent analysis suggested that oil production capacity will increase by around 18 million barrels a day by 2020 – an increase of almost a fifth in less than a decade and the sharpest rise since the 1980s.[1] But oil is only part of the picture. A recent report by Greenpeace totted up the potential emissions of fourteen major fossil fuel projects due to come on stream in the next few years. It found that they alone would be enough to push up global emissions by around a fifth by 2020. Look ahead to 2050 and the fourteen projects could have added the equivalent of around 300 gigatonnes of carbon dioxide to the air – between a half and a fifth of the total remaining global carbon budget, depending on how much risk we're prepared to take.[2]

Consciously or not, the world's fossil fuel companies and their institutional investors appear to be banking on the assumption that we will fail to solve climate change. Sometimes they're surprisingly up-front about this. Oil-giant BP, in its much-cited projections of global energy trends to 2030, doesn't even entertain the idea that we might cut carbon quickly enough to limit global warming to two degrees, considering failure a foregone conclusion.[3]

The value of oil, coal and gas reserves is the single biggest challenge to solving climate change. A global deal worth its salt would render a large proportion of those reserves worthless – at least until carbon capture technology is available. Perhaps even more significantly, it would cut the profitability of the reserves that do get burned by forcing down demand or imposing extra costs on production. To understand the scale of the challenge – and the potential write-off – it would be useful to know how much the world's fuel reserves are worth to the governments and companies that own them.

The simplest way to give a sense of scale is to consider current energy prices. At today's price of around $100, just the 1700 or so billion barrels of oil currently listed as proven reserves would add up to more than $170 trillion dollars.[4] To put that huge number into perspective, it's more than two years of global GDP, or nearly 25,000 dollars for every person currently alive. Add in the gas, the coal and the more promising-looking unconventional reserves and it's not hard to double or even triple that figure, at which point the total would stand at five to ten years of global GDP and fifty or a hundred thousand dollars for each person.

But that is a hopelessly oversimplified calculus. Much of the income from selling the existing fuel reserves would be received years or decades in the future, so a normal economic assessment would need to apply a discount rate to downgrade its value.[5] Moreover, fossil prices fluctuate, and their markets – especially in the case of oil – are distorted by cartels, subsidies and taxes. Prices have tended to go up over time and this trend is likely to continue in the absence of efforts to solve climate change, even though new technologies such as fracking for shale gas in the US have helped buck the trend for some fuels in some regions. We don't know how much higher the price paid by consumers could go before people stopped buying fossil fuels, nor how soon they may be undercut by renewables and nuclear. There's also the unconventional reserves to consider, plus the costs of extraction: building and running mines, wells and pipelines. Although almost all of the price of a barrel of Middle Eastern oil is kept as profit, margins are often much lower elsewhere.

One way to try to capture all these variables and unknowns is to use the stock market as a guide. According to campaign group Carbon Tracker, the top hundred coal companies plus the top hundred oil and gas companies listed on the world's stock exchanges had a total value of around $4.6 trillion, as of December 2012.[6] That's their net value – assets minus liabilities – according to investors. Around a tenth of this value is related to non-fuel-related activities such as mining for metals, leaving around $4 trillion as the rough current value of the listed fossil fuel sector.[7] As we'll see later, these valuations appear to be based on the widespread assumption that all the reserves can be burned and that business as usual will continue for the fuel companies.

Listed companies represent only around a quarter of the carbon in the world's proven fossil fuel reserves, however. Most of the coal and almost all the oil and gas is owned by governments.[8] If for the sake of argument we assume that investors value fossil fuel companies based on the size of their proven reserves, and that a barrel of oil or tonne of coal is worth the same no matter who owns it, that would put the total value at around $2 trillion for the coal and around $33 trillion for the oil and gas.[9] The true figures might be much lower given that private companies aren't valued entirely on their size of reserves but also on other assets such as expertise in consulting and trading. On the other hand, the true figure might be much higher given that in many cases the government-owned oil reserves are the least expensive ones to get out of the ground, and therefore the most profitable.[10]

No one can say exactly how much of this value might need to be written off by efforts to reduce emissions. It depends on everything from the future cost of carbon capture technologies to the size of the world's agreed carbon budget. But as we'll see later, some recent analysis suggests that close to half of the value of some leading oil companies may be lost, even if the world accepts coin-flip odds of exceeding two degrees and pushes hard on carbon capture. That suggests the fossil fuel assets at stake in the global climate talks are most likely worth tens of trillions of dollars to their owners. (For comparison, the amount wiped from Nasdaq companies during the dotcom crash was around five trillion.[11]) And that's not to mention all the royalties that governments may need to forgo as a result of reduced oil and gas drilling on their territory – nor the wider economic ripple effects which we'll return to in the next chapter.

Where are the fuels, and who owns them?

Considering how crucial it is to the politics of solving climate change, it's surprising that the ownership of fossil fuels hasn't been looked at more closely. Figures for the 'carbon reserves' of different nations aren't even readily available. We made our own by calculating the approximate carbon content of the proven coal, oil and gas reserves listed in the BP Statistical Review of World Energy.[12] Although this gives an incomplete picture of national carbon endowments, since most of the unconventional fuels are missed out, proven reserves give a reasonable sense of which countries could have most to lose in the short to medium term if the world imposed a serious constraint on fossil fuel use. The results are shown in the map below, in which each country has been resized to reflect its proven carbon reserves.

The level of distortion on the map shows how unevenly the world's fossil fuels are distributed. Almost three-quarters of the total are shared between just the top ten countries, despite those countries taking up less than half of the world's land area. Those ten nations are highlighted on the map and shown in the table opposite, along with the top ten measured in terms of carbon reserves per person.

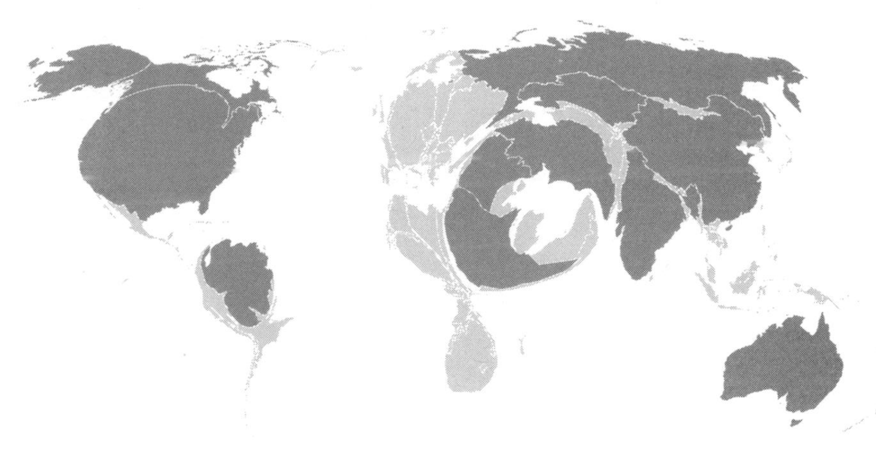

The world resized to show each country's proven 'carbon reserves': the potential carbon dioxide that would result if each country's proven fossil fuel reserves were burned. The top ten countries are highlighted in a darker shade.[13] Probable and unconventional reserves aren't included.

Total carbon reserves, top ten nations (GT CO_2)		*Carbon reserves per person, top ten nations (tonnes CO_2)*	
US	*513 GT (17.8 %)*	*Qatar*	*35,997*
Russia	*450 GT (15.6 %)*	*Kuwait*	*16,909*
China	*245 GT (8.5 %)*	*Turkmenistan*	*10,251*
Australia	*165 GT (5.7 %)*	*Australia*	*7,380*
Venezuela	*137 GT (4.8 %)*	UAE	*7,167*
Iran	*133 GT (4.6 %)*	*Kazakhstan*	*5,197*
Saudi Arabia	*128 GT (4.5 %)*	*Venezuela*	*4,745*
India	*128 GT (4.5 %)*	*Saudi Arabia*	*4,677*
Canada	*91 GT (3.2 %)*	*Libya*	*3,601*
Kazakhstan	*85 GT (2.9 %)*	*Russia*	*3,176*

For anyone who has followed the global climate negotiations over the years, one thing that jumps out, looking at these lists, is that the countries with the biggest fuel reserves have been the ones most resistant to progress in the global climate talks. The US never ratified Kyoto and has failed so far to make any really significant effort at the national level to reduce emissions. Canada ratified the treaty but then dropped out at the last minute. Russia vacillated about Kyoto and recently has tried to play the system to make money from the global carbon markets. Australia ratified only after many years of delay, despite having negotiated a target that allowed it to increase its emissions (though more recently it has upped its game, perhaps partly as a result of tangible climate changes). Saudi Arabia and the Middle Eastern countries have generally been seen as recalcitrant. China blocked targets even for developed countries in 2009, and India and Venezuela strongly resisted even the vague outline of a deal in 2011. None of this is necessarily very surprising given that, seen through the lens of fossil fuel reserves, all these countries have tangible assets that could be endangered by a major global effort to reduce emissions. Oddly, however, this relationship rarely gets discussed and has never been examined in depth.

The opposite trend also seems to be true: the countries and regions that have pushed hardest for a global deal are those without much in the way of carbon reserves. So far the strongest calls for a global deal have come from Europe, Africa and the Alliance of Small Island States. As the graph below shows, these countries control only a tiny slice of the world's fossil-fuel carbon. Other nations proactively pushing for a deal include those involved in a loose-knit coalition called the Cartagena Dialogue which includes various nations from Oceania, South America and Asia. Even if these are included, the 'yes' nations collectively control less than a fifth of the proven carbon reserves.[14] It's interesting to note that the UK, which has one of the world's most ambitious long-term emissions targets, has proven fossil fuel reserves that at current production rates would run out in just seven years (for oil), four years (for gas) and twelve years (for coal).[15] Britain's economy is also unusually reliant on imported goods and materials from China and elsewhere. These two facts together mean that the UK had relatively little to lose from setting an ambitious target.

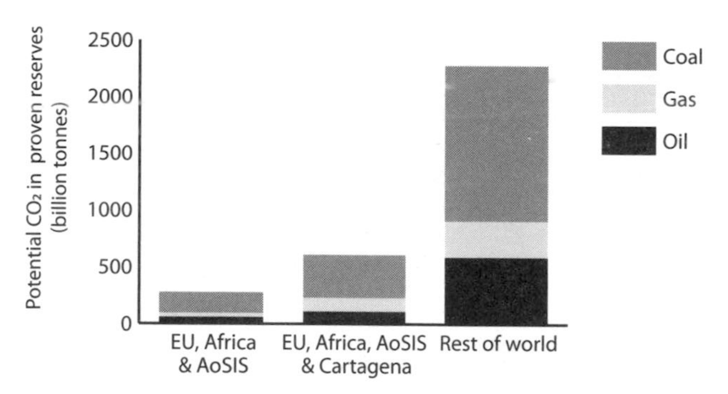

The regions pushing hardest for a global climate deal (the EU, Africa and low-lying island states) control only a tenth of the carbon in proven fossil fuel reserves. Even when the proactive Cartagena Dialogue nations are included, the sum remains around a fifth of the total.

So that's the geography, but *who* actually owns all these fossil fuel reserves? As mentioned above, big companies such as BP, Shell and Peabody own around a quarter of the proven carbon reserves. The other three-quarters are controlled by state-owned fossil fuel companies, many of which dwarf those in the private sector. These include the biggest oil companies, such as Saudi Aramco and National Iranian Oil Company; the biggest gas companies, such as Qatar's General Petroleum Corporation and Russia's Gazprom (which is listed on the stock market but with more than half of the shares retained by the state); and many of the biggest coal companies, such as China Shenhua Energy and Coal India. Governments are usually also the default owners of as-yet-unexploited fossil fuel reserves, such as oil and gas discovered under the seabed. The breakdown between corporate and government ownership varies by fuel. According to the IEA, governments own around two-thirds of the coal but as much as 90 per cent of the gas and unconventional oil.[16]

But who, in turn, owns all those corporate and government assets? In the case of listed companies, the answer is clear: the shareholders. But that isn't the end of the chain, because while some shares are held by individual investors, the majority are held by institutions such as pension funds, insurance companies and retail funds, many of them based in countries thousands of miles from the stock exchange in question.[17] Indeed, many pension funds are particularly keen on fossil fuels companies because they tend to pay reliable dividends. These institutional investors are in turn owned by or answerable to their members or customers: people with pension plans, insurance policies, and savings and investments. In other words, almost anyone with a financial stake in global society is a part-owner of a fossil fuel reserve.

The even larger government-controlled carbon reserves are – broadly speaking – also collectively owned. Certainly this is true in democracies, where the people regularly get a chance to sack and replace the administration overseeing the reserves. The ownership question is less clear-cut in countries such as China, whose huge coal stocks are controlled by unelected leaders, or Saudi Arabia, whose oil wells are owned by a royal family that rules as an absolute monarchy. Even in these countries, though, much of the income from the state's fossil fuel extraction flows back to the people through government spending or cheap energy. *Newsweek* reported in early 2012 how during the height of the Arab Spring, when commentators were wondering whether the Saudi government could be toppled, the ruling family 'allocated $130 billion in additional spending to build homes and combat youth unemployment' to help calm things down. That government revenue overwhelmingly came from oil sales.

Is there a 'carbon bubble' in the stock market?

If solving climate change will present a serious financial threat to the fossil fuel sector, that begs the question of whether oil, coal and gas companies are overvalued. Put another way, are the world's stock markets holding a 'carbon bubble'? This question was first brought to wide attention by an influential report published by Carbon Tracker in 2011.[18] It was written to highlight to policymakers and investors the growing tension between the world's stated 2°C climate goal and the market value of oil, coal and gas companies.

The report showed that just the top 200 fossil fuel companies listed on the world's stock exchanges have enough carbon in their proven reserves to generate around 750 billion tonnes of carbon dioxide – about as much as we can emit in total to give a good chance of avoiding 2°C. But we could only burn all this if the companies gave up on their probable and possible reserves – *and* if the governments of Saudi Arabia, Venezuela, China and all other countries with nationalised fossil fuel industries miraculously decided to stop all extraction from their much larger reserves tomorrow. This suggests that if we're going to succeed in solving climate change, many of the listed companies – and their investors – could be in trouble. So are fossil fuel stocks the new sub-prime mortgages? Could these toxic assets be a threat to the world's financial as well as ecological health?

Showing that there is more oil, coal and gas than can be safely burned doesn't in itself prove that there is a bubble. It could be that investors are buying fossil fuel stocks having correctly predicted that the world will never succeed in reducing emissions, or assuming that carbon capture technology will scale up and obviate much of the need for a write-off. Or it could be that they're mainly interested in short-term timeframes, in which case the flow of revenue may be more important to them than the long-term potential of a company's reserves. Some analysts claim this is especially true for coal companies.

These are possibilities, but many investors appear to have no awareness of the concept of a global carbon budget, let alone what such a budget would mean for each fossil fuel company. As one investor quoted in the Carbon Tracker report puts it: 'Valuations of the oil and gas sector still assume that they will be able to take all proven and probable reserves out of the ground and burn them.' Moreover, analysis by McKinsey and the Carbon Trust suggests that most of the value of a fossil fuel company is based on revenue that won't flow for more than a decade – by which point any successful climate deal would be significantly constraining oil, coal and gas use. A more recent report by HSBC concludes that: 'we doubt the market is pricing in the risk of a loss of value from this issue'.[19] The fact that investors are still encouraging or allowing companies to pump hundreds of billions of dollars each year into opening up *yet more* reserves appears to confirm the market failure.

One question this raises is *whose* fossil fuels we'd burn if the world succeeded in agreeing a global carbon budget. Whose assets would get written off or devalued? Oddly, this important question appears to have received very little attention so far. Common sense suggests that if we could burn only half or a quarter of the world's proven carbon reserves, we'd favour the oil and especially the gas over the coal, since they produce more energy for each unit of carbon emitted. Another reason for ditching coal first is that it's mainly used for electricity production, which is the form of energy most readily available from renewables and nuclear. (Yet, as we write this, coal companies are still floating on stock markets around the world, their IPO prospectuses entirely avoiding the question of whether their reserves can actually be burned.) On the other hand oil is mainly burned in cars, which don't last for as long as power stations, so the balance between oil and coal within a carbon constraint is hard to predict.

Whatever mix of fuels we would use in a low-carbon world, it seems reasonable to assume that we'd favour the reserves that could be most easily and inexpensively extracted. In the case of oil, that might be okay for, say, the United Arab Emirates or Saudi Arabia, which can pump crude out of the ground for a few dollars a barrel. It would probably be worse news for the countries and private corporations with higher costs of production.

A recent study from HSBC suggests that writing off 'unburnable' reserves wouldn't necessarily be a huge deal for European oil companies, because the reserves that would be abandoned would be the relatively low-value ones. BP only stands to lose around 6 per cent, according to the analysis, though some companies are more exposed, such as Statoil at 17 per cent. But that's only part of the impact; much bigger losses would be felt as lost profit margins on the more valuable reserves. How this would play out would depend on how a global climate deal was structured. In the current model – in which countries are each responsible for volunteering and hitting national carbon targets – a serious effort to solve climate change would massively reduce demand for oil imports. This would force down the global trading price, even though end users would pay much more for fuel as a result of national carbon taxes and so on.[20] The HSBC analysts believe this would present a 'material threat' to the fossil fuel sector, even with rapid development of carbon capture and storage technologies. The share of BP's value exposed to climate laws once this is considered is more than 40 per cent. Perhaps it shouldn't be surprising, in this context, that the company doesn't try to map out solving climate change as a possibility in its future energy scenarios. (Which is not to say staff there don't consider the issue; as one industry veteran said recently, there are 'no climate deniers' in major oil companies.)

However big the 'carbon bubble' really is, and whether or not it poses a significant financial threat to the world's stock markets, it's clear that fossil fuel companies have a huge amount to lose from a successful effort to tackle climate change. This 'material threat' not only helps explain climate geopolitics, but also makes sense of the efforts of fossil fuel companies to lobby against carbon laws and buy political support for their agenda. This is a particularly big deal in all-important America. According to the OpenSecrets database, individuals and PACs (political action committees) associated with the oil and gas sector donated $70 million to US candidates and political parties in 2012 alone, in addition to funding almost 795 lobbyists at a cost of more than $149 million. The coal mining sector spent another $13 million in donations and $18 million in lobbying, and the electricity utilities that consume most of that coal donated $22 million and spent a massive $145 million on lobbying. All told, those heavily invested in fossil fuels spent an order of magnitude more than those advocating alternative energy sources.[21] As we'll see in the next chapter, many fossil fuel companies have also worked hard to minimise their losses by undermining public concern about climate change.

Can we avoid the write-off by putting the carbon back into the ground?

One way to avoid writing off fossil fuel reserves while still tackling climate change is to find ways to put the carbon back into the ground after we've burned it. The standard approach to doing this is so-called carbon capture and storage (CCS), a family of emerging technologies for separating off the carbon dioxide from the exhaust stream of a power plant or industrial facility and injecting it into a permanent underground storage site, such as a saline aquifer or a disused oil or gas well. In principle, CCS might eventually be used to avoid many of the emissions from big 'point' sources, including power plants, cement and steel works. But it's not suitable for mobile or small-scale emissions sources such as cars, trucks and household boilers. In terms of fuels, it could therefore be used to capture many of the emissions of coal (which is mainly used in large plants) and some of the emissions of gas (which is burned in big power plants as well as homes and offices) but few of the emissions of oil (which is mainly used in vehicles and other small-scale equipment).

Some green campaigners oppose CCS, or at least don't actively support it, on the grounds that it's unproven, potentially dangerous, a distraction from renewables, or simply a smokescreen to justify continued fossil fuel use. These are all reasonable concerns, but the reality is that rejecting CCS can only be an option if we are confident of the prospect of leaving the fuel in the ground. Moreover, rapid development and deployment of CCS in the next few years might help to reduce the fear of a global deal among nations with the biggest coal and gas reserves. Oil-rich nations wouldn't be affected directly by this but, as we saw in chapter three, if carbon capture was cheaply available for coal and gas, we could burn quite a lot of the remaining oil in relative safety. Moreover, oil-rich nations with expertise in drilling and pipelines might be very well placed to dominate the world's CCS industries.

The various elements of CCS technology have been widely proven in other contexts. Carbon dioxide has been injected into aquifers or oil wells at hundreds of sites around the world. Paradoxically, this is usually done to help squeeze out the last drops of oil from a well, though at one site – the Sleipner gas field in Norway – the carbon is injected specifically as a result of national climate change policies. The bulk of the evidence from this and other sites suggests that liquefied carbon dioxide can be safely stored there with minimal chance of significant leakage.[22] Engineers also have plenty of experience of using chemical solvents to separate carbon dioxide from other gases, and alternative techniques – such as 'gasifying' the fuel or burning it in pure oxygen – could end up being cheaper in new coal plants.[23] That just leaves the piping infrastructure, which may be expensive but is not rocket science.

Despite the maturity of many elements of the technology and lots of small-scale trials, however, not a single commercial-scale CCS plant is yet up and running. That is as much an economic and political problem as it is an engineering one. Government support so far has been half-hearted in most countries, with flagship programmes cancelled or bungled in both the US and UK.[24] As such, although there are now dozens of pilot schemes and other trials under way, those in the industry acknowledge that we're probably still at least a decade away from any widespread deployment – and further still from CCS representing a significant part of the global energy picture.[25] Much of the problem is that the market for CCS equipment is almost entirely reliant on governments passing laws which directly or indirectly *require* power stations to fit and use it. Even if it cost nothing to install, no power station operating purely for profit would ever use CCS voluntarily as the carbon capture processes take huge amounts of energy, undermining the commercial viability of the plant. Until laws are in place to create a market, most mainstream investors will continue to see the sector as unattractively risky, despite the long-term potential for huge markets. Hence progress so far has been slow.

Of course, a global carbon deal, once agreed and enforced, would drive plenty of additional investment into CCS. With a carbon budget agreed, power companies and coal companies would have no option but to make carbon capture work or abandon their plants and reserves. But that takes us back to the Catch 22 – because without proven or rapidly developing CCS a deal will be harder to reach. Clearly there's a great opportunity here for governments that want to solve climate change to lubricate the whole issue by putting much more urgent efforts into CCS.

Even with more efforts, though, we should be under no illusion about the scale of the task required for widespread CCS deployment. To sequester just a fifth of the current fossil-fuel emissions would require an industry capable of processing and burying around 27 million cubic metres of liquefied CO_2 every day – equivalent to the total volume of fluids processed by today's oil industry, which took more than a century to scale up to its current size.[26]

Taking carbon out of the air

A more complete solution to fossil fuel write-off would be to develop ways for taking large volumes of carbon dioxide directly out of ambient air. If we could do this at an acceptable cost and with a manageable amount of energy, we could clean up after small sources such as cars and domestic appliances as well as larger industrial plants. In principle we could even reverse the clock, putting yesterday's emissions back where they came from.

Various approaches to capturing carbon from the air have been proposed. One of the simplest is to burn sustainably harvested wood or biomass (alongside coal) in conventional CCS power plants. Since each new generation of trees takes carbon out of the air, this approach would in principle make for carbon-negative power stations. A recent review of various air-capture technologies by experts at Grantham Institute at Imperial College London found that this may be one the cheapest of the options, with costs at a large scale potentially as low as $59 per tonne of carbon captured, though potentially as high as $111.[27] However, the total contribution of this approach would be limited by the amount of sustainable wood available and, of course, by the development and adoption of CCS technology.

Another approach to air capture involves rolling out machines – usually called 'carbon scrubbers' or 'artificial trees' – that use plastic polymers or other materials to react with and capture carbon dioxide directly. Since carbon mingles rapidly through the global atmosphere, the scrubbers could be placed at the best sites for long-term carbon storage, eliminating the need for new large-scale piping networks. The leading advocate of this model is Columbia University physicist Klaus Lackner, whose estimates suggest that artificial trees could take carbon dioxide out of the air for around $100 per tonne – and perhaps eventually much less.[28] So far, however, the various companies working on air capture technology have struggled to attract serious funding and no such device has been demonstrated at scale.[29]

A completely different proposed route for getting carbon out of the air is to use farming techniques that accelerate the build-up of organic matter in the world's soil. The top metre of soil around the planet already contains more carbon in organic form than the atmosphere contains as carbon dioxide, despite centuries of soil degradation and desertification caused by agriculture, building and wildlife loss. If soil carbon levels could be boosted by just a tiny increment each year, therefore, that could mop up a very significant amount of carbon dioxide. Various approaches promise to be able to help with this. These include zero-tillage farming (growing crops without ploughing the land) and the burial of charcoal or 'biochar' produced from crop residues or wood. In both cases, there's evidence that these techniques can sequester carbon – potentially in large quantities – while also improving soil fertility. The soil-carbon approach most rapidly gaining interest, however, is the idea of restoring the world's huge degraded arid lands using special cattle-grazing techniques designed to mimic the environmental impacts of herd animals such as bison and wildebeest.[30] The key thinker in this field is Allan Savory, who has spent decades seeking to reverse desertification by introducing livestock to degraded soils. The idea is to ensure that animals move around in the manner of a wild herd being pursued by natural predators, grazing the land to keep it healthy but quickly moving on before it gets over-exploited. This regenerates the land with the side effect of turning it into a giant carbon pump.[31]

Not everyone is yet convinced that soil-based approaches could scale up enough to make a difference to the need to cut fossil fuel use. In a 2009 review of geoengineering options, the UK's Royal Society concluded that soil-based carbon capture techniques 'may be useful ... on a small-scale although the circumstances under which they are economically viable and socially and ecologically sustainable remain to be determined.'[32] However, Savory believes that techniques already demonstrated by his land-management institute – if rolled out widely enough – could return the atmosphere to its preindustrial carbon levels at the same time as massively increasing global food production.[33] He claims that achieving a 'reasonably easy' two per cent increase in soil organic matter 'over the bulk of the world's rangelands' would both tackle desertification and sequester 2880 gigatonnes of carbon dioxide equivalent – roughly comparable to the world's proven fossil fuel reserves.[34] The claim that such a remarkable feat could be easily achieved isn't yet backed by any major independent scientific reports, and questions remain about the likely amount of planet-warming methane generated by billions of additional livestock. But if Savory is even partly right then his scheme could make a significant difference to the chance of inexpensively solving climate change.

More research and development funding is urgently needed into all the various forms of carbon capture, high-tech and low-tech alike. If advocates such as Lackner and Savory are proved right, their solutions could prove pivotal to unlocking the whole issue and minimising the fossil fuel write-off. It's abundantly clear, however, that the world can't afford to hold its breath and wait for carbon capture to save the day. As the Grantham Institute report concludes, such techniques are essential to develop but 'should not be used as an excuse for delaying effective global mitigation efforts'.

The infrastructure write-off

The value of oil, coal and gas reserves is a huge deal, but there is more. We also have a substantial investment in all the power stations, furnaces, cars, lorries, tractors, planes, boilers and other equipment that we've built to burn all those fuels. Infrastructure naturally becomes obsolete over time but at the moment we're building it faster than the old stuff is wearing out. The result, as we'll show, is that soon it may be impossible to stay within a safe carbon budget without forcing lots of infrastructure into early retirement – or at least reducing its value by making it more expensive to use.

Infrastructure is important because it gives companies, governments and citizens an economic incentive to resist climate change legislation. The more money they've sunk in, the more they stand to lose. A company that owns a coal power plant could face bankruptcy if carbon taxes make it less profitable; a government that owns oil rigs has value at stake if some of its reserves can't be profitably extracted; and a couple who have just invested in an expensive car would be more likely to reject a government threatening to drive up fuel prices, effectively reducing the value of their asset by making it more expensive to use.

As with the fossil fuel reserves, we can't say for sure how much all the carbon-burning infrastructure is worth, but we can get a rough sense of scale. In its *World Energy Outlook 2011*, the IEA forecasts that $17 trillion of investment will be needed in the power generation sector by 2035. That must put the value of today's total power-sector infrastructure well into the tens of trillions of dollars. The world's motor vehicles must be of a similar order of magnitude given that the global automotive industry turns over more than $2 trillion every year and there are more than a billion registered vehicles worldwide.[35] It looks as if current infrastructure assets are therefore less valuable than the reserves themselves but not necessarily by a huge margin.

How much of this value may eventually need to be written off depends on when we reach the point where we have enough infrastructure to exceed a safe carbon budget in its normal lifespan. The IEA explored this question and estimated that existing infrastructure – including vehicles, buildings and the energy sector – already 'locks us in' to burning four-fifths of the carbon permitted by 2035 to give a 50 per cent chance of avoiding 2°C.[36] Another study by academics at Stanford reached a similar conclusion.[37] At first glance these results look quite positive, because they suggest that if we stopped building any new plants, vehicles and other devices that require fossil fuels to run, and just used the ones that already exist or are under construction until the end of their normal lifespans, we'd most likely be okay.

The problem, of course, is that we're still building more. In fact, we're building more faster than ever. At current rates, the IEA predicts, as soon as 2017 we'll have built all the kit we need to exceed our coin-flip budget to 2035. Once we reach that point, if we're to stay within our budget, every new combustion engine or car or pipeline or minehead or truck or boiler produced will require another one of a similar size to be retired early. That will add significantly to the costs of avoiding climate change. The IEA estimates that each $1 that we fail to invest in transitioning to low-carbon energy systems in the next decade will eventually cost us $4.3 further down the line – adding up to hundreds of billions of dollars of additional costs and, more importantly, creating more economic resistance to climate legislation.

So how is the world responding to this conundrum? By shutting its eyes, it seems. A report by the World Resources Institute in late 2012 found that at least 1,200 coal power stations are currently in planning. The majority of these are due to be installed in China and India but plants are also planned in more than fifty other countries. By our calculations, if all these plants are constructed and used until the end of their working lives, they alone would 'lock in' around a third of the remaining all-time carbon budget compatable with a good chance of not exceeding 2°C.[38] Vehicles with fuel-burning engines are also multiplying rapidly. In 2011, the world produced around 80 million cars, vans and trucks that rely on petrol or diesel to run – a massive 42 per cent increase compared to ten years before – and there's no sign of a slowdown.[39] Chemical works, boilers, tractors and other fossil fuel infrastructure appear to be on a similar upwards trajectory. Add in all the infrastructure that already exists and it's clear that the write-off here may need to be huge unless the world rapidly changes course or finds reliable ways to get carbon out of the air.

Three levels of lock-in

Large-scale infrastructure such as power plants and blast furnaces are the key to reducing carbon lock-in – both because they account for the majority of emissions and because they last so long. When a new petrol-burning car rolls off the forecourt, it will typically stay on the road for ten to twenty years; when a new coal power plant is opened, it will usually stay in operation for thirty to forty years.[40] For that reason, it's crucial that the world does whatever it can in the short-term to limit the rate at which these long-lived fossil-fuel plants are put up. Each new one erected will increase emissions and create future corporate lobbyists pushing against progress on climate change in order to avoid their paymasters having to write-off their assets.

But vehicles matter too, not least because cars are where most of the world's voters have their main investment in fossil fuel infrastructure. Decades of experience have shown that any rapid increase in fuel prices – as would almost certainly have to result from a binding global climate deal – can be politically very difficult for governments. Hence the more rapidly the world can tighten up fuel efficiency standards for new vehicles, the more political space it will open up to reduce current and future fossil fuel use.

Another reason vehicles are important is that there's a whole second level of infrastructure required for their manufacture. That includes the factories, the component manufacturers and even the plants that produce the machines that go into the factories. These aren't considered in the IEA's analysis, but they matter. Automotive companies today are still planning and building new plants for producing petrol and diesel cars. To avoid making a loss, these may need to produce cars for decades; and the cars in turn will last for another decade or two on top of that. All of which will lead either to increased emissions or additional write-off.

Nor does infrastructure lock-in stop at factories. Yet another layer exists in the built environment – and the legacy here is even longer-term. Every new housing development built beyond walking distance from local amenities and every expansion of road infrastructure is predicated on the assumption of abundant energy for private transport. Every new airport assumes a rise in the use of aviation fuel. All this extra energy could one day come from carbon-free sources – renewable electricity for cars, say, or algae-based biofuels for planes – but that will only add to the already immense challenge of meeting our energy needs in a low-carbon way. This is an issue where the lock-in lasts and lasts, because most homes, offices, airports and other buildings are designed to function well into the next century. But most governments are ignoring this and permitting or commissioning construction projects that will add yet more resilience to the carbon curve. China alone is planning *seventy* airports over the next five years, along with the introduction of more than 2,000 new planes.[41]

Discarding and adapting infrastructure

Although the rate at which we're throwing down fossil fuel infrastructure is a huge cause for concern, the term 'lock-in' does deserve a brief health warning, because it implies that if we've got the infrastructure we are forced to use it until it is no longer functional. In reality, adaptation may be practical in many cases, whether that means retrofitting carbon capture to coal power stations or fuelling conventional cars with cellulosic ethanol made from sustainably harvested wood – or even liquid fuel made from air, water and renewable power.[42] And if that fails there is still the possibility of voluntarily retiring stuff early. That is costly but not unusual. A visit to any municipal dump reveals how readily society throws out old TVs, computers and other pieces of kit that it regards as no longer desirable, well before the end of its working life. On a bigger scale, China recently closed a number of small coal power stations to replace them with larger, more efficient alternatives that wouldn't pose as many risks to local air quality.

There's also the possibility of adapting infrastructure. The Second World War saw factories around the globe making radically different products at very short notice. So perhaps car factories built to produce conventional cars could retool and start producing electric ones instead – or even carbon capture devices.

Indeed, in some cases it may be possible to put fossil fuel infrastructure to a completely different use as dwindling availability of oil, coal and gas starts to render them uneconomic. This kind of thing has also been done before. In 1981, the giant oil-fired Bankside power plant on the Thames in central London shut down, driven out of the market by rising fossil fuel prices. We now know that building as the Tate Modern – the most visited modern art gallery in the world. Similarly, in the heartlands of the industrial revolution, Manchester's 'dark satanic mills' now make desirable offices and flats. The High Line in Manhattan is two miles of redundant overhead rail track transformed into an urban park. Many of the world's canals tell a similar story.

Nonetheless, fossil fuel infrastructure *is* a crucial issue, as the more money global society sinks into vehicles, plants and buildings that require oil, coal and gas to function, the more resistance there will be to constricting supplies of those fuels. One key to solving this problem is a clear signal from policymakers that carbon *is* going to be cut, starting now. But many different players can help too. Financial analysts can highlight the risks of investing in fuel-reliant infrastructure. Investors can heed their advice. Campaigners can push for significantly higher emissions standards for new plants and vehicles so that future infrastructure has a lower level of carbon lock-in. Planners and governments can insist on a built environment that is fit for a low-carbon world. And citizens can demand it of them.

9. The growth debate

Would solving climate change hit GDP – and does it matter?

Owners of fossil fuel reserves and infrastructure may have the most obvious stake in ongoing oil, coal and gas use, but the fuels are only worth so much because the world's people and organisations create a market for them. Consumers, companies and public bodies around the globe spend trillions of dollars on fossil fuels each year. Some do so to support energy-profligate lifestyles – complete with the gas guzzlers, patio heaters and frequent flier miles of carbon footprint cliché. But that's a relatively small part of the picture. As the table overleaf shows, fossil fuels also power our factories, hospitals, commutes, schools, farms, trucks – even the internet. Almost everything we do in society carries a carbon footprint.

In that context, and in a world where most people – and most journalists – spend more time worrying about rising fuel prices than they do about rising carbon emissions, it's no wonder that policymakers are nervous of doing anything that may reduce fossil fuel availability. Even if the interests of oil, coal and gas companies could be overcome, both affluence and quality of life may be at risk – and in turn the power of those currently in office.

So what *would* happen to our economies and lifestyles if governments did an ambitious global carbon deal? Would GDP tumble if we rapidly constrained fossil fuel use, as happened in the 1970s oil shocks, and as some policymakers and environmentalists seem to assume? Or, as advocates of 'green growth' claim, could the economy keep growing as fossil fuels give way to low-carbon energy sources and rapidly rising energy efficiency? And if growth would be endangered, how much would that affect wellbeing? These questions attract polarised and often deeply entrenched views. Many people take growth to be an unquestionable cornerstone of a healthy and happy society. Others see the very concept as lying at the heart of the world's climate crisis and want to see it purged from policymaking.

Sector	Footprint (kg CO_2e/person)
Services – 27%	
Government, defence, etc.	830
Financial, legal, professional	240
Restaurants, pubs & hotels	600
Construction	920
Health services	430
Education	250
Other services	500
Driving, including car production – 15%	
Driving to work	419
Driving to school or college	96
Driving to see friends	458
Driving to the shops	276
Driving for holidays/leisure	490
Other driving	322
Shopping – 12%	
DIY & gardening products	480
Appliances & gadgets	270
Newspapers, books, paper products	140
Home furniture & furnishing	190
Clothes, shoes & bags	200
Other stuff	370
Home electricity – 9%	
For computers, phones, etc.	128
For TVs, hi-fis, gadgets, etc.	199
For water heating	173
For laundry	111
For fridges & freezers	168
For heating	149
For cookers, toasters, etc.	73
For lighting	186
For dishwashers	44
Home gas or oil use – 13%	
Water heating	512
Central heating	1222
Gas cooker	36
Leisure flights – 8%	
Short-haul leisure flights	451
Long-haul leisure flights	716
Public transport – 3%	
Train travel	68
Ferries & boats	125
Coaches, buses & taxis	192
Water & sewage – 2%	
Home water supply	40
Home sewage	279

The UK's total carbon footprint broken down by sectors and expressed in kilograms of carbon dioxide equivalent per person. The figures include all regulated greenhouse gases and take account of imported and exported goods (though not deforestation).[1]

Limits to growth?

The discussion of climate change and economic growth often gets bogged down in a broader and longer-standing environmental discussion about resource scarcity. Green thinkers have tended to argue over the decades that economic growth will eventually have to stop as energy and resources start to run out – a concept popularised in the 1972 Club of Rome book *Limits to Growth*. According to this school of thought, the whole idea of an economic system that assumes 'infinite growth on a finite planet' is inherently flawed and the inner workings of capitalism will need to be reinvented to create a sustainable economy.

Sceptics of that view argue there are plenty of resources left for the foreseeable future and that with better and more efficient technology powered by effectively unlimited supplies of renewables or nuclear energy, we could grow the economy for decades or centuries to come, recycling finite materials where constraints exist or even taking them from space. (A company, backed by the founders of Google among others, was recently set up to explore the possibility of mining metals from asteroids and bringing them back to Earth.[2]) With climate change, however, natural resource limits aren't the problem. Quite the opposite, in fact: we have way more fossil fuel than we can burn, so the relevant question is not whether we'll run out but whether growth is compatible with deliberately cutting its use.

This is a hugely important question because our political and economic systems eat and breathe the language of growth. GDP is the world's most powerful metric, the dominant benchmark by which governments are judged by citizens, the media and each other. Policymakers the world over tell us that while they're keen to cut carbon emissions and protect the planet, they won't do so at the expense of a growing economy. President Obama made just this point when he finally broke a long silence on climate change in November 2012: his administration would cut emissions as long as it didn't endanger growth.[3]

We'll return to the question of whether GDP is a suitable measure for human progress later, but given how entrenched we are in pursuing its growth, we had better ask whether that goal would be endangered by the kind of deal needed to tackle climate change.

How would tackling or not tackling climate change affect growth?

Ever since the invention of money, the global economy has grown almost continuously, and as we saw in chapter six this growth has tended to move in lockstep with energy use. This begs two important questions. First, would it be possible to keep increasing the global energy supply while cutting emissions fast enough to deal with climate change? Second, if growth in energy use is steadily slowed or stopped by our carbon cutting efforts, can we break the age-old link and keep the economy growing regardless?

The short answer to both of these questions is that no one can say for sure. Economists can run models and make predictions, but, as with modelling the future climate, there are many layers of uncertainty. For a start, we don't know how much cheaper renewable energy, nuclear power and CCS might eventually become, nor how rapidly they could scale up. We don't know what future technologies may evolve to produce, save, store or transport energy. (As the internet revolution has shown, technology can sometimes evolve beyond recognition in little more than a decade.) Crucially, if alternative energy sources *don't* scale up in time, we don't know how sensitive our economies would prove to be to a steady reduction in total energy supply. Neither do we know how high fuel prices would go within nations if the global supply was constrained, nor *whose* energy consumption would change and in what way. On top of all this, we have no idea how many trillions of dollars of value would crash out of the stock market as a result of downgrades of fossil fuel companies, nor the extent to which the value of other sectors would rise to compensate.

Published in 2006, the *Stern Review* was the first major attempt to look at the costs both of solving and not solving climate change. Commissioned by the British government from Sir Nicolas Stern, former chief economist of the World Bank, it is still the most prominent report of its kind, though its conclusions were highly contentious right from the outset. It drew on a range of models from various research bodies to estimate how global GDP might be affected by efforts to cut carbon emissions.[4] Their average prediction was that tackling climate change would lead to GDP being just 1 per cent lower in 2050 than it would be with neither climate change nor any attempt to solve it. A couple of years later Stern said he believed 2 per cent was more realistic given rising emissions and the growing urgency of the science, but such a reduction would still be relatively small, equating to a loss of annual economic growth of a tiny fraction of 1 per cent.

Many of Stern's assumptions about the costs of and resistance to carbon cutting were optimistic, however. He didn't look at fossil fuel write-offs and since the report was published emissions have accelerated rather than slowed down. Hence the carbon cuts now required to avoid 2°C now look much tougher. A detailed paper by Kevin Anderson and Alice Bows in 2011 found that to provide coin-flip odds of limiting temperature rise to 2°C, global emissions would need to be cut by 5–6 per cent every year from 2016 in rich countries and from 2020 in developing countries.[5] Cuts of that speed are unprecedented except in times of recession and have rarely been seriously considered by policymakers or mainstream analysts. In this context, some commentators, such as Anderson and Bows themselves, have concluded that 'dangerous climate change can only be avoided if economic growth is exchanged, at least temporarily, for a period of planned austerity'. Other commentators such as economist Tim Jackson agree, rightly pointing out that so far there is virtually no historical evidence that the world is capable of decoupling carbon from growth.[6]

But the evidence is not clear cut. The world has never *tried* to phase out fossil fuels before. And there are plenty of more upbeat assessments from centres such as the IEA and International Institute for Applied Systems Analysis which suggest that with enough political will it would be possible to increase efficiency and roll out renewables, nuclear, biofuels and CCS fast enough to replace fossil fuels at the rate required. It wouldn't be cheap but neither would it be impossibly expensive once you factor in the money saved by not building new fossil fuel infrastructure. The IEA estimates that the scenario shown in the graph below, designed to meet anticipated future energy demand while giving a coin-flip chance of limiting warming to 2°C, would require

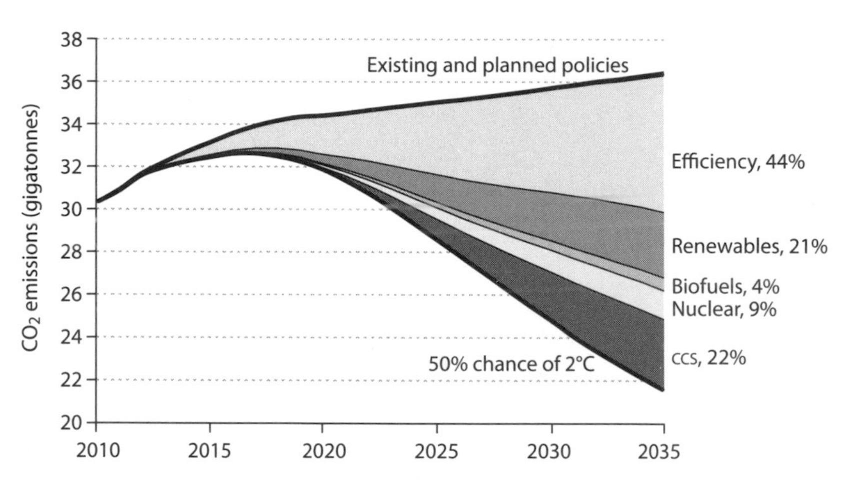

A scenario modelled by the International Energy Agency showing what technologies and efficiencies might be needed to cut emissions by the level required (the bottom line), relative to policies already announced (the top line).

additional global investment in efficiency and energy generation of around $160 billion per year in the current decade, rising to $1.1 trillion annually by 2035.[7] These are large sums but not enough in themselves to take down the global economy, especially given that the money would help stimulate new industries.

These kinds of scenarios wouldn't be plausible without some kind of global deal to constrain fossil fuel use, due to the rebounds and other balloon-squeezing effects explored in part two of this book. But would they be plausible *with* a global deal? This is an open question. It depends not only on the technical and economic uncertainties listed earlier, but also on how smartly the world's green policies are designed – and on how readily we embrace them. Onshore wind is a good example of how crucial public opinion can be. It's the cheapest form of clean energy in much of the world, but local communities often block turbines in their area on aesthetic grounds. Similarly, many people reject nuclear energy or CCS for a host of understandable reasons but, as we'll see later, many energy experts believe that rejecting either would make the job much harder.

For all these reasons, we'll only know what effect an ambitious global deal would have on the economy once we give it a go. If the world acts fast enough and policymakers win public backing, it looks possible that uninterrupted global economic growth might be compatible with cutting carbon at the required rate. It's even feasible that massive government investments in low-carbon infrastructure could help reinvigorate the ailing economy in many countries. After all, this kind of thing has worked in the past – most famously in F. D. Roosevelt's New Deal, the massive public building programme that helped cut unemployment and stimulate the American economy after the great crash of 1929. But while a 'green new deal' is an attractive idea that deserves to be enthusiastically pursued, it remains highly plausible that writing off trillions of dollars in the fossil fuel sector and reducing emissions by more than 5 per cent a year globally – and much faster in developed countries – would significantly slow down growth or lead to an economic contraction in the short term, especially if society is slow to embrace alternative energy sources. It's sobering in this context to remember that almost all the major global recessions of the last fifty years were preceded by spikes in oil prices.

Even if the economy rode the storm well at the global level, nations heavily reliant on selling or burning fossil fuels might find their economies shrinking due to lost export revenues or rising production costs. In the absence of widespread public backing for a global deal, this could spell a serious political risk for rulers of carbon-rich nations such as Russia, Australia, China, the US and Saudi Arabia.

Of course, the risks to GDP of rapidly phasing out fossil fuels need to be compared to the risk of *not* acting – though once again, the uncertainties are vast. The *Stern Review* estimated that unchecked warming would eventually reduce global GDP by around 5 to 20 per cent. Such figures are inherently problematic, however, as Stern acknowledges when he writes they should be treated with 'caution and humility'. After all, it's perfectly plausible that runaway global warming could devastate human society and wipe out most of the world's species. How can any economic model meaningfully put a price on that kind of scenario, especially given that no one can quantify the precise likelihood of it coming to pass?

Even if we could accurately predict and cost the future impacts of catastrophic climate change, there's the subjective but important question of how much we should 'discount' those costs to reflect the fact that we usually care more about the present than the future. On one level, applying a discount rate is a matter of common sense. But as a policy decision this has major philosophical implications; it's akin to deciding how much we value ourselves over our children and our grandchildren.[8]

How crucial or desirable is GDP growth?

Whether or not we can have it while phasing out fossil fuel use, the extent to which we actually *need* GDP growth is another question that attracts polarised and entrenched views. Plenty has been written about GDP's shortcoming as a measure of progress. Even within the terms of conventional economics, it is a crude indicator of material living standards, thanks to its focus on production rather than household consumption, and its failure to reflect income distribution. As a broader measure of human wellbeing, it's even more profoundly flawed. A huge body of research has shown that once average incomes rise above a certain level continued increases correlate neither with the average person's stated life satisfaction nor with objective wellbeing indicators such as life expectancy, trust, obesity, family stability and murder rates.[9]

In poorer countries, rising affluence *does* tend to go hand-in-hand with increased health and happiness, which isn't surprising considering that it usually accompanies the roll-out of basic infrastructure and social services. But in all countries there's another fundamental limitation with focusing on short-term GDP: the possibility that today's gains may be tomorrow's loss. As green growth advocate Michael Jacob puts it: 'When economies run down fish stocks, or farm so intensively that soils are degraded, or cut down forests without replacing the trees … the income earned is counted towards GDP, but the loss of the "natural capital" that can sustain this income into the future is ignored.' In the case of global warming, the natural capital in question is a hospitable climate in which human society can thrive.

For both these reasons, it's clear that short-term GDP is a profoundly inadequate measure of human progress. And if it's true, as management gurus have long argued, that 'what gets measured gets done',[10] it seems likely that there are opportunities to improve both our wellbeing and our sustainability by focusing on better metrics that relate more strongly to our current and future quality of life. While there's no reason to assume that a low-carbon society would score higher on wellbeing in the short-term than a fossil fuel society, at least with better metrics we'd be chasing what matters. They make sense whatever we do about climate change.

The country pushing this idea most strongly is the small Himalayan kingdom of Bhutan, which has been getting noticed by the world's media in recent years with an alternative metric, Gross National Happiness, designed to reflect 'the peace and happiness of our people and the security and sovereignty of the nation.'[11] Some Western policymakers have made cautious moves in a similar direction. In 2008, Nicholas Sarkozy, then president of France, commissioned a major report on alternative economic indicators from a panel of leading economists, chaired by Nobel prize-winner Joseph Stiglitz. The commission concluded that 'the time is ripe for our measurement system to shift emphasis from measuring economic production to measuring people's wellbeing'.[12] In Britain, Prime Minister David Cameron instructed the Office for National Statistics (ONS)to increase its efforts to measure wellbeing in order to keep more meaningful tabs on 'how the country is doing'.[13] In the academic sphere, mainstream economists are also exploring new metrics, such 'national balance sheets setting out assets and liabilities', designed to ensure that each generation passes on as much income-generating potential to future generations as we currently enjoy.[14]

This is all promising talk and could help pave the way for societies that are both happier and less nervous that taking action on climate change might slow down GDP. But new measures like this need to be taken seriously. They need the power that GDP has had – and so far they're simply not getting it. In Britain, for example, in the same month that the ONS published its first detailed wellbeing survey, the prime minister who commissioned it announced that he wanted every government department to become 'a growth department'. To show he meant business, he replaced his transport minister to clear the path for a new runway at Heathrow airport (a U-turn on a previous environment pledge) and had his finance minister, along with those of the other G7 nations, call on OPEC to pump more oil into the global economy.[15]

For all the talk of alternative metrics, then, short-term GDP remains the pre-eminent political priority, even in wealthy countries where its benefits aren't clear cut. But why? Part of the answer is the simple fact that it's deeply engrained into the fabric of our institutions, from treasuries and boardrooms to newsrooms. But there's also the fact, too often ignored or played down by critics of GDP, that while more growth doesn't always lead to greater wellbeing, the *absence* of economic growth can cause all sorts of problems, such as high unemployment and home repossessions, with much of the pain shouldered by society's poorest and most vulnerable. This is not only problematic in itself but a major political barrier for any government considering a policy that could endanger economic growth. Whether it's a democratic administration seeking another term in office or an unelected government aiming to retain its power base, history shows that a steadily growing economy will almost always increase the chances of success.

Adding to the political nervousness is the knowledge that, even if growth continues unabated, the public may resist even relatively small lifestyle changes that might be incurred by a serious response to climate change. Cutting emissions might require us to accept electric cars over petrol ones, for example, or to put up with much more expensive flights. We might need to tolerate more wind farms, more nuclear plants, or both – or allow insulation to be installed in our leaky homes to make them more energy efficient, leading to aesthetic changes either inside or out. None of these are terrifying in themselves but policymakers know how change-resistant society can be. Just look at the furore in some parts of the media when inefficient light bulbs were phased out – something that almost certainly boosted the economy but still met fierce, almost obsessive resistance.

The combined fear of fossil fuel write-offs, economic fall out and modest cultural change has led to a state of political paralysis in many countries. Even those policymakers who in principle want to take serious action on climate change are terrified of the short-term potential backlash. There's only one way around this kind of paralysis: public pressure.

10. The great global slumber

Why are we not more fired up about climate change?

Given the political and economic barriers to tackling climate change, a solution will remain unlikely without the world's citizens being properly alert to the problem and pushing hard for action. But thus far popular concern about climate change has been conspicuous by its absence. A small proportion of people, deeply worried about global warming, are making efforts to call for change or lead by example, but most us, from punters to presidents, act as if we just don't *care* very much – even though, for the most part, we now accept the science.

This chapter looks briefly at the psychological and social barriers that are holding us back. These are surely the most overlooked and critical dimensions of the climate change problem, which is all too often viewed solely in terms of science, technology, politics and economics. If the world felt fired up enough about the issue, citizens would shift their behaviour and voting to match their concerns – even taking to the streets if necessary. Greener governments would feel mandated to make difficult decisions. Fossil fuel lobbyists would still complain, but they couldn't compete with widespread public pressure.

So what is it about us, our societies and the nature of the problem that has caused us, collectively, to sleepwalk into so much danger?

An unbelievable problem

If you wanted to invent a problem to induce confusion, disbelief and the turning of blind eyes, it would be hard to come up with something better than climate change. It's caused by a build-up of gases that we can't see, smell or taste and the effects play out through a weather and climatic system that is by its nature unpredictable and variable. Although we're already seeing impacts, no single flood, drought or storm can be definitively attributed to climate change. It takes academics with supercomputers to work out how much the risks have changed and even they can't say for sure since the details of the science are so complex and contain so much uncertainty.

Adding to the abstract nature of the problem is the fact that the most dangerous impacts are many years away. By the time we see climate changes shocking enough to act as a global wake-up call, we will be committed to many decades of worsening symptoms and it may be too late to stop runaway warming. It's like mushroom poisoning: by the time you know you're ill, the drug is ingested so far into your system that there may be nothing you can do. Except that poisoning is simple to understand, historically proven and an immediate threat to us as individuals. Man-made climate change, by contrast, is complex, unprecedented and its worst impacts will be felt by people we won't meet, decades into the future.

This mix of abstraction, complexity and long-term uncertainty provides the human mind with all the wiggle room it needs for avoiding or playing down the uncomfortable facts of the situation – and in some ways we might be innately predisposed to doing just that. We like to see ourselves as rational creatures, but we're often ready to buy into whatever is most comfortable or enjoyable to believe. We like ideas that fit with our existing worldview. We like to believe what our friends believe. And most of us like to believe that the future looks rosy. Psychologists such as Nobel laureate Daniel Kahneman have been documenting these kinds of systematic 'biases' in our thinking for decades.[1]

One example is our tendency for wishful thinking. Neuroscientist Tali Sharot, author of *The Optimism Bias*, argues that 'a growing body of scientific evidence points to the conclusion that optimism may be hardwired by evolution into the human brain'.[2] Supporting this claim is a wealth of research that shows that people tend to wildly underestimate their chances of experiencing negative outcomes – such as divorce or cancer – and significantly overestimate their chances of winning the lottery, having gifted children, achieving professional eminence or living to a ripe old age. This sunny outlook may help to lower stress and improve physical health, Sharot and others argue, but it also makes us prone to sitting back and hoping for the best when such a response doesn't make sense.

Reinforcing our naturally rose-tinted view of the future is the knowledge that previous environmental concerns have usually gone away without much fuss. Governments fixed the ozone layer and acid rain. Atomic technology didn't lead to a global nuclear holocaust as green and peace campaigners used to fear. And thanks to innovations in farming and science, Thomas Malthus was wrong when he predicted that food production would never be able to keep up with population growth. None of these is a good parallel with climate change, but the sense that previous crises were averted or exaggerated nonetheless adds to our sense that global warming will somehow work itself out. (Indeed, many of the people who once claimed that climate change didn't exist now prefer to brand those concerned about the issue as neo-Malthusian scare-mongers.)

Another psychological barrier to dealing with climate change is our tendency for short-termism. When Abraham Lincoln quipped that 'the best thing about the future is that it comes one day at a time' he touched on a truth about human nature: we find it much easier to consider today and tomorrow than we do to think months, years or decades ahead. Some evolutionary biologists argue that this short-termism is just as hard-wired as our optimism bias. Richard Dawkins writes that, 'The only solution to the problem of sustainability is long-term foresight, and long-term foresight is something that Darwinian natural selection does not have.'[3] This needn't be an insuperable barrier: even Dawkins acknowledges that the human brain 'is well able to override its ultimate programming', but that will require additional effort and creative thinking. One ingenious idea mooted recently by the World Future Council is to appoint 'ombudspersons for future generations' – officials at the national or international level whose role is to represent and fight the corner of people who don't yet exist or are too young to get taken seriously.[4]

Perhaps the most serious psychological barrier to tackling climate change, however, is the human facility for interpreting facts in a way that supports our values, prejudices and the expectation of our social groups. This trait – sometimes described as 'confirmation bias' – underpins much of the deepest resistance to climate change science. For example, it's convenient for people with libertarian political perspectives to see global warming as not being dangerous as it avoids acknowledging the need for government regulation. Similarly, it's helpful for consumers to believe that climate change will be resolved at the political level as that view avoids confronting the ethical implications of our decision to keep on flying, driving and shopping as usual.

Confirmation bias works on all sides of the debate. Climate change is an attractive concept for people who, for whatever reason, are excited by the idea of modern capitalism imploding or mother nature putting humans back in their place. But that too is unhelpful as it means that some individuals waxing lyrical about climate change do so with a hint of glee or as part of a broader political narrative (not to mention with frequent factual errors). This serves as further proof to those with opposite instincts, causing them to become yet more entrenched in their positions. And so a debate which demands to be had with open-mindedness has been stifled by the polarisation of attitudes and even more biased and selective interpretations of evidence on all sides. One recent study showed how people's cultural and political values determine even things such as whether or not we notice or acknowledge changes in local weather patterns.[5] As economist J. K. Galbraith put it, 'faced with the choice of changing one's mind and with proving there is no need to do so, almost everyone gets busy on the proof'.[6]

Even when we aren't dodging the facts, the abstraction of climate change makes it hard for us to latch onto emotionally. From Chilean miners to child abductions, the world's media and citizens can be held rapt by stories of specific individuals facing understandable threats, even when the people involved are unknown to us. But climate change will impact on almost everyone, and in varied and complex ways, so it fails to tug the heartstrings. (No wonder online campaigning groups often find they get a much lower rate of engagement with emails about climate change than they do about more emotive subjects with clearly defined villains and victims.) We also struggle to feel much emotion about our own contribution to the problem. To kill or injure someone in front of our eyes is abhorrent. Most of us drive carefully because we understand what a car crash is like. The imagery triggers an emotional response, our foot slackens on the accelerator and good habits are reinforced. But to detract fractionally from billions of current and future lives is just a vague concept – especially if it results from the release of an invisible gas emitted through the tangled supply chains of all the goods and services we buy.

Summing up these various psychological barriers, Anthony Leiserowitz of Yale University writes of climate change: 'You almost couldn't design a problem that is a worse fit with our underlying psychology'.[7] These barriers are not only a major problem in themselves; they also leave the door wide open for dirty tricks.

Sabotage

With tens of trillions of dollars' worth of fossil fuel reserves and infrastructure at stake, it's perhaps not surprising that some of the businesses and individuals with the most money to lose from addressing climate change have done whatever they can to block progress. For decades, many of those invested in ongoing oil, coal and gas use have pumped money into lobby groups, think tanks and PR agencies with the express purpose of persuading people and politicians – especially in America – that climate change either doesn't exist, doesn't matter or will be impossibly expensive to solve. This sabotage has taken many different forms: television and billboard advertising, attempts to influence classroom teaching, even 'astroturfing', where PR professionals pose as disinterested punters while leaving comments under online news articles and blog posts. Almost always, though, the messaging is designed to exploit our psychological soft spots.

Aware that we're bad at dealing with uncertainty, the fossil fuel lobby has worked especially hard to discredit the science that links greenhouse gases to global warming. As Naomi Oreskes and Erik Conway document in their book *Merchants of Doubt*, this approach is identical to those employed to undermine the links between passive smoking and lung cancer, industrial pollution and acid rain, and CFCs and the ozone layer.[8] Remarkably, the same handful of fervently anti-regulation scientists who have campaigned against action on climate change were also centrally important to these other campaigns. In each case their contrarian (and usually non-specialist) perspectives were amplified by a powerful network of corporate-funded think tanks and lobby groups.

In the 1990s, energy and automotive companies openly sought to block climate legislation via bodies such as the Global Climate Coalition. This group was so successful at shifting the debate that the Bush administration credited it with playing a key role in America's rejection of the Kyoto Protocol.[9] When the organisation eventually shut in 2002, it was able to say quite credibly that it had 'served its purpose'.[10] Not to leave anything to chance, however, many fossil fuel companies – most notoriously ExxonMobil – continued to fund think tanks that could do their work at arm's length. These included outfits such as the Competitive Enterprise Institute, which in 2006 released a high-production-value TV ad campaign which claimed that glaciers were growing, not shrinking, and said of carbon dioxide, 'They call it pollution. We call it life.'[11] More recently, the Heartland Institute (funded partly by the owners of the US petroleum corporation, Koch Industries) went so far as to associate people who believe in climate change with mass murderers. Its billboard campaign showed mug-shots of serial killers alongside the words: 'I still believe in global warming. Do you?' Heartland's president, Joseph Bast, said on the accompanying press release, 'The most prominent advocates of global warming aren't scientists. They are Charles Manson, a mass murderer; Fidel Castro, a tyrant; and Ted Kaczynski, the Unabomber. Global warming alarmists include Osama bin Laden and James J. Lee.'[12]

Decades of public-facing campaigns such as these have doubtless done plenty of damage to public understanding of global warming, but their effect has been small compared to the way in which fossil fuel interests have co-opted deep-seated political divisions. Perhaps because politicians on the right tend to be more anti-regulation and pro-business, they're usually the most natural and willing political friends of fossil fuel companies. But as oil and coal money has flooded into rightwing parties, a feedback loop has emerged: campaign contributions have led to stronger political rhetoric against climate change, which in turn has led to a greater sense among conservative voters that 'people like me' reject climate change. This in turn strengthens the political rhetoric, boosting the attitude polarisation described earlier and increasing the social barriers to people changing their minds.

The money, politics and arguments have become so bound up that the US Republican party has often proactively done climate doubt-mongering on behalf of its corporate sponsors. In a now notorious 2003 memo to President George W. Bush, strategy consultant Frank Luntz wrote: 'There is still a window of opportunity to challenge the science … Should the public come to believe that the scientific issues are settled, their views about global warming will change accordingly. Therefore, you need to continue to make the lack of scientific certainty a primary issue in the debate.'[13] Some politicians have gone much further. Senator James Inhofe, who sits on and previously chaired the Senate Committee on Environment and Public Works, has described global warming as a hoax and compared environmentalists to Nazis. At the last count he had taken campaign contributions from those in the oil and gas industries totalling $1,414,996.[14]

The blurring of climate and politics is significantly strengthened by parts of the press aligned with the politicians and voters in question. Much of the most direct distortion of the facts comes in the shock-jock media. In the US, that includes Fox News, home to commentators such as Sean Hannity ('The debate's over; there's no global warming') and Glenn Beck ('A discredited global warming scam').[15] In the UK, the *Daily Express* recently dedicated a whole front page to a list of '100 reasons why climate change is natural.'[16] Even parts of the up-market media regularly publish misleading material, however, whether that's the editorial page of the *Wall Street Journal*, Christopher Booker's column in the *Sunday Telegraph*, or documentaries such as Channel 4's much-criticised *Great Global Warming Swindle*. A recent analysis by the US Union of Concerned Scientists found that Fox News was scientifically inaccurate on climate change 93 per cent of the time, while the *Wall Street Journal* opinion page was misleading 81 per cent of the time.[17]

The mainstream media is also surprisingly negative towards the technologies that could replace fossil fuels. One recent study looked at a month's coverage in the most popular UK broadsheet and tabloid newspapers and found that more than half of the articles covering the renewables industry were negative – and only 28 per cent positive.[18] Similarly, the hit car show *Top Gear* has *twice* been caught faking a scenario where an electric car runs out of power during a review or feature to make for a more entertaining storyline.[19]

Not content with support from think tanks, politicians and the media, the fossil fuel industry continues to pedal its own message, too. A *New York Times* analysis found that in the first half of 2012, a presidential election year, nearly two dozen groups spent more than $153 million on 'television ads promoting coal and more oil and gas drilling or criticising clean energy'. That worked out as 'nearly four times the $41 million spent by clean-energy advocates, the Obama campaign and Democratic groups to defend the president's energy record or raise concerns about global warming and air pollution'.[20] The adverts promoting fossil fuels usually combine a common sense tone with emotive themes such as the need for national energy security. One American Petroleum Institute advertisement opined recently that 'New energy taxes could hurt drivers and families ... Better to produce more energy here, like oil and natural gas. That will help the economy. That's good for everyone.'

A majority concern

Despite the sabotage, and the abstract nature of the climate change threat, popular concern is inching upwards, and the numbers who reject the science are in decline, even in America. There are, of course, ups and downs in the polls on global warming. For example, concern dipped in many countries after the 2009 climate conference in Copenhagen failed to achieve a deal and a cache of leaked emails appeared to implicate some climate scientists in underhand practices.[21] But overall, global polls from the last couple of years suggest a significant and growing majority of people *are* concerned about global warming.

In 2010, an Ipsos MORI poll found that almost half of the world's people ranked climate change as one of the top two or three challenges facing the world, putting it joint first with war/terrorism and above economy and poverty.[22] A Pew survey from the same year showed that more than 70 per cent of the people in India and South Korea, and 91 per cent in China, were happy in principle to pay more for energy in order to cut carbon emissions.[23] In the biggest survey yet on people's perceptions of the risks of climate change, carried out in 2012, a remarkable 97 per cent of respondents in emerging economies and 81 per cent of people in developed nations say they are 'worried' or 'very worried' about climate change.[24]

Traditionally, the averages in the global polls have been brought down significantly by the US, where climate scepticism is much more common – probably thanks to the vast flows of fossil fuel money in politics there and the resulting perception of global warming as a left–right issue. In the Ipsos MORI poll mentioned above, 67 per cent of Chinese people ranked climate change as a leading global issue, but only 22 per cent in America. This is massively important because the chance of an ambitious global climate deal without serious American participation is extremely low: the developing world will quite reasonably refuse to constrain fossil fuel use without serious efforts from the US.

As we write this chapter, however, American acceptance of global warming is on the rise – a trend that appears to have been driven by the climate itself.[25] In 2012, the US experienced an extraordinarily hot spring in which weather records were not just broken but broken by record-breaking margins. Then came a devastating drought in the Midwest that wiped out vast areas of crops, driving up food prices. In parallel, the summer-time sea ice in the Arctic collapsed at unprecedented speed to reach a new record low weeks earlier than scientists expected. All of this is consistent with global warming – and all of it seems to have set a ball rolling with public opinion. A *Washington Post* survey in the summer of 2012 showed that 78 per cent of Americans think 'global warming will be a serious problem' if no action is taken, with 55–61 per cent saying the government and businesses should do 'a great deal' or 'quite a bit' about it.[26] Another poll showed that 65 per cent of Americans support 'imposing mandatory controls on carbon dioxide emissions/ other greenhouse gases.'[27] And a survey in the autumn carried out by Yale University found that 74 per cent of Americans think 'global warming is affecting weather in the United States' – up five points since March.[28] And all of that was *before* Hurricane Sandy brought unprecedented devastation to New York and the eastern seaboard just days before the presidential election, catapulting climate change onto the political agenda for the first time in years and causing *Bloomberg Business Week* to declare on its front cover, 'It's global warming, stupid.'[29] With increasingly tangible impacts, the polls suggest we're slowly starting to wake up to the threat of climate change. By and large, though, it remains an 'armchair concern': most of us still aren't demanding that politicians or companies take action or indeed cutting our own carbon footprints. So what's holding us back?

Social inertia

Part of the answer, perhaps, is that humans are social creatures and the ways we act are significantly determined by what everyone else is doing. As social psychologists like to point out, when a fire alarm goes off, we don't look for smoke to determine whether to evacuate; instead, we look to see whether anyone else is running for the door. (One remarkable experiment showed that even when there *is* billowing smoke, people take ages to respond if their peers aren't already doing so.[30]) So when most people, politicians, businesses and others aren't doing anything significant about climate change, we feel a natural reluctance to be the first to jump – even if we think it's a good idea in principle.

This social inertia plays out at many different levels. On the micro scale, few of us are keen to be the family member who suggests cut back on flying or the colleague who suggests boycotting a useful supplier on the grounds that it has been named and shamed for lobbying against progress on climate change. Taking such stances leads to social tension that we could do without. The risks are even larger in the all-important communities where climate change science is rejected. As one researcher put it recently:

> Take a barber in a rural town in South Carolina. Is it a good idea for him to implore his customers to sign a petition urging Congress to take action on climate change? No. If he does, he will find himself out of a job, just as his former congressman, Bob Inglis, did when he himself proposed such action. Positions on climate change have come to signify the kind of person one is. People whose beliefs are at odds with those of the people with whom they share their basic cultural commitments risk being labelled as weird and obnoxious in the eyes of those on whom they depend for social and financial support.[31]

The barrier of social norms also plays out through the media. Research by psychologists Amos Tversky and Daniel Kahneman showed people tend to assign undue relevance and weight to issues they are exposed to a lot.[32] They found, unsurprisingly, that people 'tend to assess the relative importance of issues by the ease with which they are retrieved from memory – and this is largely determined by the extent of coverage in the media'. At the moment, most of us hear relatively little about climate change, even if our choice of media accepts the mainstream science. One recent study compared American media coverage of reality TV show *The Kardashians* with media coverage of ocean acidification caused by carbon emissions, which threatens to radically alter marine life with huge implications for life on earth. *The Kardashians* received fifty times more coverage in US newspapers and 270 times more coverage on television news. Editors would presumably justify this by arguing that they're just reflecting the interests of their audience, but therein lies a vicious circle because the media coverage unquestionably helps determine what the audience is interested in.

A similar vicious circle of responsibility plays out between voters and policymakers. Although most individuals believe that climate change should be solved by government and corporations, when policymakers or companies are asked why they're not doing more to tackle the problem, a typical response is that voters and customers aren't demanding it.

Ultimately, the only way to break through the social inertia on climate change is for some people – politicians, punters, celebrities, business leaders, journalists, barbers, shoppers, voters, anyone – to seize the initiative and show leadership, whether that means taking political action, cutting carbon or simply standing up for the facts. After all, pack behaviour works both ways: the more people who *do* start talking about or taking action on climate change, the more others may follow. Anyone taking a leadership role may therefore have a disproportionate effect – especially those in communities where there's a social pressure to pretend the issue doesn't exist.

Is consumer culture a factor?

Many green thinkers and other commentators believe that one reason we're so reluctant to reduce fossil fuel use and get engaged with issues such as climate change is that our culture has become so focused on consumption. This not only drives up our carbon emissions, the argument goes, but also stops us engaging the big issues. Economist Tim Jackson sums up this view, writing that 'our relentless search for novelty and social status locks us into an iron cage of consumerism.'[33]

A strong desire to consume wouldn't matter to global emissions if the world agreed and implemented a carbon budget. With such a deal in place, each of us could seek to consume as much as possible, safe in the knowledge that while our own consumption might deprive someone else, it wouldn't change the world's carbon footprint. But unless more people start demanding political action or making a statement by cutting their footprints and expecting it of others it's unlikely that such a deal will ever be agreed. So it's worth asking what drives our consumer-focused culture.

Whether or not the desire to consume is partly innate there's no doubt that in various ways modern culture pushes us along – most obviously through advertising. Across society, appeals for greener lifestyles and political engagement with climate change are mere whispers compared to the endless, sometimes deafening, messages pushing the idea that we'll have more fun, be more respected, and have better friendships and relationships if we have more, travel more, use more and even eat more. We are given many of these messages implicitly, and without regard to the truth.

The evidence that advertising directly increases our consumption and carbon emissions as opposed to redirecting our spending from one product to another, is 'inconclusive' according to a recent report on the subject by the Public Interest Research Centre.[34] But common sense suggests the barrage of adverts promoting things such as inefficient cars or long-haul short breaks simply must be unhelpful. Even if it doesn't drive up our footprints directly, advertising reinforces cultural values based on image and status that could make people less likely to engage with a big collective challenge such as climate change. Some senior people in the industry are surprisingly up-front about this. Rory Sutherland, speaking as President of the Institute of Practitioners in Advertising, recently wrote that: 'I am much keener that we should accept the vast moral implications of what we all do and debate them openly rather than fudge the issue.'[35] That surely is a discussion worth having.

Inequality has also been suggested as a possible driver of apathy and consumerism. In *The Spirit Level: Why Equality is Better for Everyone*, Richard Wilkinson and Kate Pickett argue that in developed nations the gap between rich and poor makes everyone more desperate to reach the top of the pile and to prove to all, through their possessions and activities, that they've got there. This status anxiety not only decreases wellbeing in all manner of ways, the book claims, but also drives our carbon footprints. In the chart below we've tested this theory by plotting the inequality figures from the *Spirit Level* against the average carbon footprint of people in each country. Unlike with most of the social trends explored in the book, there's no obvious link here. That's not necessarily very surprising, because while it's possible to see how inequality could drive the *desire*

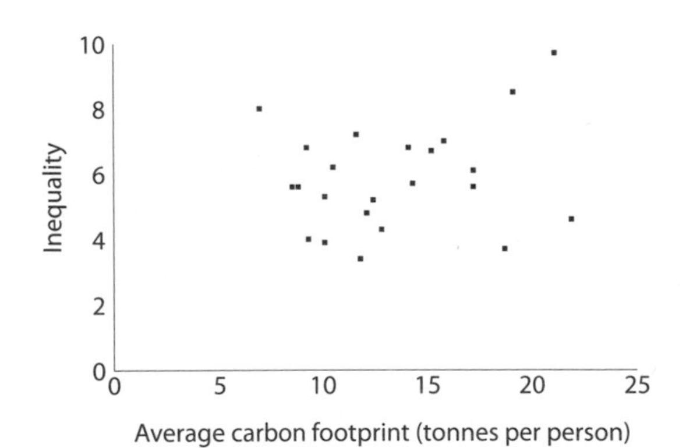

Inequality plotted against average total carbon footprint in OECD nations, represented by the dots. There is no statistically significant trend between the two.[36] But inequality may be a barrier to tackling climate change in other ways.

to consume, it's also possible to see how it could be a brake on consumption overall. A billion additional dollars in the pockets of the average person may be more likely to enable additional home heating or holiday flights than they would in the pocket of the super-rich who already don't worry about heating bills and take as many holidays as they like. There is some academic research to support this idea.[37]

Overwhelmingly, carbon footprints are driven by affluence and energy mix, not other factors such as social inequality. But the *Spirit Level* authors point to a second reason why tackling inequality might help solve climate change and here the argument seems stronger. Their suggestion is that more equal societies tend to be more 'public-spirited' and trusting – and this could make those nations more amenable to a global solution to climate change, even if it doesn't encourage them to cut their carbon unilaterally. There appears, for example, to be a strong correlation across developed counties between social equity and recycling rates; another piece of analysis suggests that business leaders in more equal countries 'are more strongly in favour of their governments complying with international environmental agreements than business leaders in more unequal countries'.[38] Whether equality leads to public-spiritedness or public-spiritedness leads to equality isn't clear, but the link is interesting and the relationship makes intuitive sense: the more divided we feel by inequality, the less likely we'll feel inspired to work together to solve big collective problems such as climate change. This is true not only between individuals at the national level, but even more so between countries at the international level. Which takes us neatly to the fourth major barrier to solving climate change.

11. The problem of sharing

A tragedy of the atmospheric commons

In 1968, the controversial ecologist Garrett Hardin coined the phrase 'tragedy of the commons' to describe the challenge of managing a limited resource that everyone has unlimited access to. His example was a communal pasture land. As the population rises and the herds grow, all the farmers might agree that the land is starting to become degraded due to overgrazing, putting everyone's livelihood at risk. But it's not in any individual's economic self-interest to remove their cattle if no one else does.[1] One way around this problem is to divide up the land into private patches, enabling farmers to manage their own areas. (This, rightly or wrongly, is exactly what has happened to most of the world's communal grazing over the centuries.) The alternative is to agree some kind of system to determine who has access and when.

Climate change poses a similar challenge. Although most of the world's citizens and nations have agreed that we're degrading the global atmosphere by dumping too much exhaust gas into it, few of us want to be the first to stop. The difference, it has often been pointed out, is that the atmosphere by its nature can't be fenced.

The tragedy of the atmospheric commons plays out at the micro and macro levels. If as individuals we avoid a flight to reduce our carbon footprint, or spend an hour writing to a politician, it doesn't mean that we personally – or our friends and relatives – will experience noticeable climate benefit. Instead, any benefit will be unquantifiable and dispersed over more than seven billion people, the vast majority of whom we will never meet and who will never even know that we exist. From an individualistic standpoint, it doesn't make much sense. As writer

Chris Goodall points out, it's likely that getting rich would be a more efficient way for any person to protect their own offspring from global warming than trying to persuade the world to give up on fossil fuels.[2] Hence the only strong reason to act is the sense that it's the right thing to do – either in our own eyes or in those of people whose esteem we hope to win.

At the national level we deal with these kinds of 'bigger-than-self' problems with government regulation. Most of us pay our taxes – however grudgingly – even if others get most of the benefits. We usually have no choice, of course, though the knowlege that everyone is subject to the same rules sweetens the pill. But with climate change precisely the same challenge exists at the international level. While we can define and defend national airspace from foreign planes, there's no way to stop the intermingling of carbon emissions over national borders. Austrian emissions affect Argentina and vice versa. In the absence of atmospheric fences, that only leaves the alternative approach to solving the problem: agreeing a system to determine which nations can extract or emit what, and when. Given that the world's multi-trillion-dollar energy system is at stake, and countries have profoundly different levels of wealth, fossil fuels and historical emissions, this was always bound to be tricky.

How not to share a commons

No matter how the world tries to negotiate a global climate deal, burden sharing will remain a key challenge. But the current approach seems particularly unlikely to overcome that challenge. The main problem is that there's no structure. Each country or region simply makes its best offer for how much it's prepared to reduce its emissions and is then responsible for administering these cuts by whatever means it likes. Options include increasing taxes on fossil fuels and subsidising the alternatives, or implementing a cap-and-trade scheme in which a limited number of tradable permits for the production or use of carbon-based fuels are distributed to companies by auction or some other means. Whichever policies are chosen, though, the result will be rising fuel prices and this presents a major political challenge – especially if the rest of the world isn't doing the same.

With enough public pressure, perhaps this disorganised process could lead to the level of carbon cuts needed. But even if it did, an approach based on arbitrary national targets is prone to those targets being dropped or revised in individual countries with a change of government or situation – as Canada showed when it dropped out of the Kyoto Protocol. For both these reasons, it might be more sensible to scrap the current approach and start again by first agreeing a global carbon budget and then negotiating a single coherent system for implementing it – such as a worldwide cap-and-trade scheme or a global carbon tax system.

With global agreement, such a scheme could be relatively simple to design and run. It could be administered either at the point where fossil fuels flow into the world economy (such as oil refineries, coal mines or ports) or at the point of consumption (petrol pumps, electricity bills, and so on). In theory the two should have the same effect as the oil refineries and coal mines would just pass on the cost to end consumers. But as various thinkers have pointed out over the years, it would be much more practical to regulate carbon as close to the point of extraction as possible. After all, there are far fewer oil refineries than petrol stations and homes.[3]

But a globally administered carbon tax – or a cap-and-trade scheme in which carbon permits are sold at auction – would raise huge sums of money, begging the question of what should be done with it all. One option would be to simply distribute the money back to countries in relation to their populations. In practice this would mean countries with high carbon footprints would end up worse off, and countries with low carbon footprints would end up better off. But poorer nations might demand a larger slice of the tax takings to reflect the fact that they have released less carbon historically and have minimal capacity to roll out new energy systems. Similarly, nations that produce and export fossil fuels – such as Saudi Arabia or Russia – may demand a disproportionate share, since they stand to lose the most. It might be possible to find a compromise but the negotiations would be messy. Quickly we'd be back to the same chaotic squabbling that's failed to deliver progress on climate change for the past two decades. Furthermore, in countries where there's a strong vein of anti-government sentiment, such as the US, there would likely be plenty of resistance to paying tax that goes to the government of, say, Venezuela.

Simplifying

If the idea of a rules-based system is good, but the prospect of sharing out the tax or auction revenue is offputting, another approach would be to allocate emissions permits to nations based on some kind of simple formula. The most popular proposal along these lines is 'contraction and convergence', in which a steadily falling number of permits would be allocated between nations each year. At first, countries would get enough permits to cover their current emissions, but each year the total would decline and the allocation would become more equitable, so that by a fixed date in the future – say, 2030 – countries would receive permits in proportion to their populations. Permits could be traded, so India, for instance, could choose to sell permits to Australia rather than using its full allocation directly. The idea is simple, transparent and fair, and it reduces the number of variables in the negotiations to just two: how fast should global emissions come down and how quickly should nations converge on an equitable per-capita share.

An even simpler model is a proposal called 'SAFE Carbon', developed by a group of academics including climate scientist Myles Allen of Oxford University.[4] This system does away with taxes and permits all together. Instead, anyone taking carbon out of the ground in the form of coal, gas or oil would be required to pay for some carbon to be put back into the ground. A company extracting oil in Iran, say, or gas in Russia might pay a coal power station in the US or China to sequester the required amount of carbon on its behalf. That amount would gradually rise, from zero per cent of the carbon extracted at the start of the scheme to a hundred per cent at the point where the world's budget has been used up. Framing a deal this way would help accelerate investment into CCS, Allen argues, simultaneously increasing the feasibility of steep emissions cuts and quickly opening up a huge global market for carbon sequestration. This in turn might sweeten the deal for countries with major oil and gas industries, as their expertise in pipelines, wells and chemical engineering would give them a head start in that huge new market.

Neither of these proposals is a silver bullet for the burden sharing problem, however. Poor nations would almost certainly object to SAFE Carbon as it would simply drive up global fossil fuel prices, which would be much easier to accommodate in rich nations than in developing ones. By contrast, rich nations with high per capita carbon emissions might object to C&C because (if warming is to be limited to 2°C) it would require them either to make *massively* steep carbon cuts or become reliant on buying huge quantities of valuable permits from nations they see as competitors, such as populous China.

Give and take

Ultimately, no matter how a deal is structured, not everyone can win. Even with a rapid roll-out of carbon capture, total fossil fuel use will need to be constrained in the coming decades, and if all nations try to get the rights to extract or burn as much of the available fuel as possible, a deal will remain elusive and emissions will continue to rise, at a massive long-term cost to everyone.

The only viable solution to this conundrum is for large numbers of people to make it clear to their governments that they care more about solving climate change than they do about getting the best possible deal for their nation. Given the rate of carbon cuts required, success will require all nations to make concessions but especially those in the rich world with its far greater levels of affluence and historical responsibility.

Some will argue that humans are just too selfish and nationalistic to make such national concessions – and that the bulk of people in wealthy countries will never be able to muster enough empathy for faraway people and places to support a deal that disadvantages them at home. If they're right, then we had better start preparing for major climate disruption, because without some give and take on the world stage, the chance of rapidly bringing down global emissions is going to be much lower.

But not everyone is so pessimistic. Empathy and global-mindedness are cultural variables, after all. Neuroscientists such as Susan Greenfield argue we can strengthen synapses for such traits in almost the same way as we do when learning a language or musical instrument.[5] The darkest episodes of human history also point to the variability of empathy levels, albeit in the wrong direction. If our concern for others can collapse, does that suggest it could also be somehow encouraged to increase?

PART FOUR

Not just fossil fuels

The other ways we're warming the planet

12. The supporting cast

Agents of warming, fast and slow

So far we've looked at climate change solely as an energy issue. But fossil fuels aren't the only source of carbon dioxide and carbon dioxide isn't the only cause of warming. We now turn to the other ways that human activity is raising the temperature. These contribute enormously to the problem and offer some of the quickest wins for reducing the impacts. Considering these other gases and particles brings food and land properly into the picture and also sharpens our focus on a few specifics such as aviation and refrigeration whose footprints are only partly down to energy consumption.

In the case of carbon dioxide, the two key sources after fossil fuels are the clearance of tropical forests – to provide farmland, timber and access to minerals – and the production of cement for the world's construction industry. These sources have helped sustain the long-term exponential path of the carbon curve from chapter one, and need to be accounted for in any global carbon budget. The more effectively we can tackle them, the less daunting the task will be on the fossil fuel front.

In addition to all the carbon dioxide, we release a host of other gases and particles that amplify the greenhouse effect. Two of the most important are methane and nitrous oxide, generated in large quantities by farms, mines, gas rigs, landfill sites and sewage works. Many refrigerant gases are also powerful greenhouse gases and leak into the air from industrial plants, fridges, freezers and air conditioners. Also significant is the soot, or more accurately 'black carbon', generated not just by industry and vehicles but by millions of cooking fires and stoves across the developing world. (Soot also lands on glaciers, darkening their surface and speeding up their rate of melting by enabling them

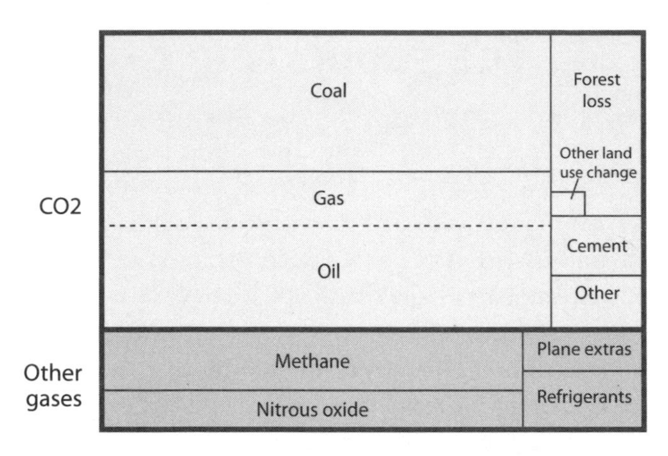

Today's global greenhouse gas emissions weighted to show their relative warming over a century.[1] Fossil fuels account for two-thirds of the carbon dioxide and 72 per cent of the total. The separate cement and aviation areas show the warming impacts of those sectors over and above their fossil fuel use. Each would double in size if its slice of oil, coal and gas was included. Soot isn't included here as its impacts are not well understood and it's not regulated under current global warming legislation, but it's hugely important – probably second only to carbon dioxide in terms of current warming.[2]

to absorb more heat from the sun.) Even the vapour trails – or contrails – left in the sky by aeroplanes warm the planet by trapping heat that would otherwise escape to space.

Each of these agents warms the climate at a different rate. Nitrous oxide and some refrigerants are similar to carbon dioxide in that they can stay in the air for centuries, causing warming for as long as they remain there. The others – sometimes dubbed short-lived climate pollutants or SLCPs – do their damage more quickly. Methane causes a more vigorous warming effect than carbon dioxide but has a half life of just seven years.[3] It also leads to the formation of ground-level ozone, which has an even more powerful impact but stays in the air for just days or weeks. Soot, contrails and some refrigerants are equally short lived.

To complicate the picture yet further, the world's coal power plants, shipping fleets and various other sources throw out large quantities of particulate matter: tiny particles that lead to the formation of aerosols in the atmosphere. Aerosols reflect sunlight back into space (both directly and by stimulating brighter clouds to form), resulting in a powerful but short-lived cooling effect often described as 'global dimming'. This effect isn't perfectly understood, but it's thought to have reduced the amount of sunlight reaching the earth's surface by a few per cent.[4] For now, this masks the full effect of the greenhouse gases we've added to the air. As the world starts rapidly to phase out coal (as it must to tackle climate change) and continues to tighten up local air pollution standards, global dimming's cooling effect will be diminished, counteracting some of the gains.

Candles and blow torches

Because of the differing timeframes over which the various gases and particles have their effect, their relative importance depends a great deal on how far into the future we look. To help make sense of this, imagine the earth as a cauldron of water. Emitting carbon dioxide is akin to adding candles underneath, whereas releasing shorter-lived climate pollutants is like applying a short blast with a blow torch. After a minute, the candles haven't made much difference but the blow torch has had its full impact. After a couple of hours the cauldron might have been equally warmed by the two treatments, in which case the candles will have had a bigger impact by the end of the night.

So it is with global warming. If our aim was to reduce the impacts that we will experience in twenty years' time, stopping today's carbon dioxide emissions would look relatively unimportant, because the gas does its 'damage' slowly and after a couple of decades its effect will have been relatively slight. (The benefits might even get cancelled out in the short term by a reduction in global dimming as coal pollution disappears.) By contrast, over the same timeframe, reducing methane, soot or aviation emissions would seem more beneficial, because two decades are long enough for them to have had almost all their impact. But look further into the future and everything changes, because today's carbon dioxide will still be there in the atmosphere causing warming in decades and centuries to come, whereas today's methane and soot emissions will have largely been and gone.

The most frequently adopted convention for assessing the relative importance of the different warming agents is to measure their 'global warming potential' over the space of a century. From this perspective carbon dioxide dominates, as shown in the diagram earlier in this chapter. But the rationale for the hundred-year convention is not particularly strong. We've already discussed real risks of 4°C temperature rises well before the end of this century and there's a strong risk of hitting 2°C within fifty years.[5] Moreover, people are already suffering from more extreme weather events as a result of today's elevated temperature. If we shift to a fifty-year view to reflect this, the fast-acting agents would double in significance relative to the others and the importance of addressing their sources would correspondingly double. According to one study, efforts to cut soot and methane alone could hold down the temperature in 2050 by almost half a degree – far more than would be possible in that timeframe by cutting slow-acting carbon dioxide.[6] Given that this gain could be achieved at relatively little cost and with significant positive side effects, this is an opportunity that the world can't afford not to take.

But cutting the fast-acting agents without also tackling the carbon dioxide (and nitrous oxide) would just be pushing the problem further into the future. Indeed, the thing that will ultimately determine whether the climate crosses a dangerous tipping point is the maximum temperature that human activity

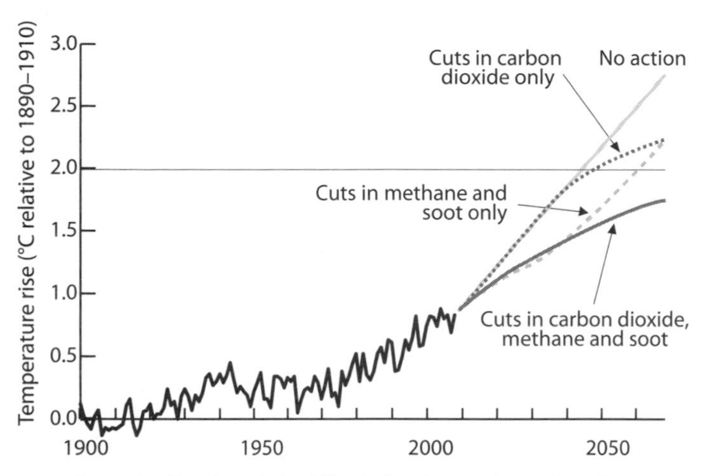

Prompt action to cut short-lived climate agents, such as the methane and soot modelled here, is the key to reducing the rate of global warming in the coming decades. But until carbon dioxide emissions are also falling fast, such efforts will only delay rather than reduce the all-time temperature peak.

pushes it to. Because that will take many decades or centuries to play out, its level will be determined principally by the accumulation of carbon dioxide in the air. Indeed, detailed modelling shows that unless we rapidly bring down carbon dioxide emissions, tackling the short-lived agents can only delay, not reduce, the all-time temperature peak.[7] A delay would still be much better than nothing, since an extra decade or two might allow us to develop technologies for sucking carbon dioxide out of the air, just in time to prevent runaway feedbacks from kicking in. But the modelling results show that cutting the fast-acting gases can't be a substitute for dealing with fossil fuels and other sources of carbon dioxide.

The obvious conclusion from all this is that the world needs to deal with the long-lived and short-lived gases and particles: the candles *and* the blow torches.

Quick fixes

Thankfully, there is considerable scope for reducing the short-lived agents and the other sources of carbon dioxide. One key area for action is the global food sector, as we'll see in the next chapter. But there may be even quicker wins in other areas. A massive roll-out of inexpensive cooking stoves could reduce soot emissions and ease demand for firewood at the same time as reducing the air pollution that kills millions of people from respiratory and lung disease each year in the developing world. The most potent refrigerant gases – HFCS – could be phased out at low cost and there's already a push for this to be incorporated into the Montreal Protocol, the treaty that has successfully dealt with the depletion of the ozone layer. A much higher proportion of the methane emitted by landfill and water-treatment sites could be captured and burned for energy. There are promising technologies for low-carbon cement, too – including one based on magnesium oxide which absorbs carbon dioxide from the air as it dries, making the process carbon negative.[8] Rolling these kinds of technologies and solutions out globally would be a big challenge but compared with getting off fossil fuels, it doesn't look too intimidating.

Aviation – a high-altitude problem

A more problematic area that deserves a special mention is aviation. Even though the sector currently uses only a tiny slice of the world's fossil fuels, its total warming impact is significantly higher due to vapour trails and the other effects of burning fuel at altitude. Although impossible to pin down precisely, this is thought to roughly double its impact to around 3 per cent of total greenhouse gas emissions on a century timescale, or a much larger proportion over a shorter timeframe. More importantly, aviation emissions are growing fast – at around 6 per cent per year – and the world seems particularly committed to this trend as society globalises ever further and the emerging middle classes realise travel ambitions.[9]

Unfortunately, flights look like being one of the hardest activities to turn green. Alternatives to liquid fuels have barely got off the starting blocks and while biofuels could offer an alternative to kerosene, powering the world's planes this way would place huge pressure on farmland, as we'll see in the next chapter. There's some potential for making carbon-neutral liquid fuels from renewable energy and water, but it's not clear that will be commercially viable.[10] Moreover, alternative fuels do nothing to deal with the vapour trails and other high-altitude impacts.

Efficiency gains could offer marginal improvements. Flight-path improvements could save a few per cent – or even more if we moved away from manual air-traffic control systems and allowed computers rather than humans to choreograph the world's planes, just as they do for modern rail networks.[11] Better payload optimisation would also help, and optimising the length of hops could shave off another small slice by avoiding planes carrying more fuel than necessary. In the longer run, new aircraft designs could further increase efficiency, but there is a physical limit to how much this can help.[12] Moreover, safety legislation ensures that new designs take years to come through – and even when they do, that won't in itself write off the current fleet, much of which looks set still to be in service in twenty or thirty years' time.

With growth massively outstripping efficiency gains for the foreseeable future and no breakthrough technologies on the horizon, it seems inevitable that tackling climate change would cause plane ticket prices to soar. This could turn out to be something of a touch-point for resistance to climate laws both in rich and emerging countries. (As we'll see later, even the EU's modest effort to limit the carbon footprint of global air travel led to fierce resistance and threats of trade wars from other countries and regions.) Not that fewer flights would be the end of the world. Many business trips could easily be substituted by video conferencing and as prices rise much air-freight would simply lose out to shipping and local production. But in the UK at least, the vast majority of the aviation emissions stem from the quest for sun, sea and sand, which can only be reduced by people flying less often or less far.[13] How much of any global carbon budget we use for flying therefore depends to a large extent on the depth of our devotion to international travel, but it's possible to imagine aviation's current 3 per cent growing to become a quarter or more of total emissions by mid-century.

13. Food, forests and fuels

Can we feed the world while cutting emissions?

Climate change, food and farming are closely interlinked. Food production is a key driver of global warming, accounting for around a quarter of total greenhouse gas emissions – and the majority of the emissions that aren't caused by fossil fuels. But food is also an area of human vulnerability to climate change because rising temperatures could make it much tougher for us to grow what we need. Even our attempts to *solve* climate change have knock-on effects on the food system by increasing the incentive to turn crops into biofuels.

When thinking about food, however, there's obviously more than just climate change to think about. Roughly a billion people – one in seven – currently goes hungry.[1] This happens today, even before the population swells by two or even three billion over the coming decades. We also know that, by driving deforestation and giving over ever more of the earth's surface to intensive agriculture, the way in which we currently manage our food supply is hammering the world's biodiversity. The Earth is currently losing species at around a hundred or a thousand times faster than the normal background rate and some scientists believe this poses 'a greater threat than global climate change to the stability and prosperous future of mankind on Earth'.[2]

When thinking about the food and land system, all this needs taking into account. It's a triple challenge: feed everyone, cut emissions and protect biodiversity. Oh, and we like it to taste good, too. This last detail, trivial as it must seem to those who don't have enough, drives the industry at least as hard as the need for nutrition itself.

Food's impact on the climate

It may sound unlikely that the food system could generate a quarter of global greenhouse gas emissions but that's the rough total if you tot up everything involved in the production, transport and retail of food. For a start, farms and food networks use plenty of oil, coal and gas. A global carbon cap as discussed in the last part of the book would take care of these carbon dioxide emissions, albeit with some impact on food prices. But agriculture also impacts on the climate in three other ways. It's closely bound up with deforestation, since most of the land cleared by destroying tropical forest is eventually used for farming or grazing. Agriculture is also the main source of nitrous oxide, which is generated in huge quantities by fertilisers, and the principle source of methane, which is belched out by ruminating animals such as cows and sheep and emitted from manure and flooded paddy fields.

In addition to all this – and too often ignored – is the fact that the global food system currently puts a lot of pressure on the world's soils. Intensive farming and overgrazing can both lead to reductions in soil's organic content with the result that more carbon dixoide ends up in the air. The climate impact of this is hard to measure precisely but is potentially hugely significant – as is anything that can reverse the process.

Deforestation and the land crunch

Recent years have given some cause for cautious optimism around deforestation.[3] While our use of fossil fuels has continued to soar in the last decades, emissions from deforestation have fallen substantially, both in real terms and as a proportion of the total. Deforestation used to account for more than 20 per cent of carbon dioxide emissions, but it's currently somewhere around the 10 per cent mark.[4] Unfortunately, as we saw in chapter one, the global carbon curve has been unaffected by this as more fossil fuels have come on-stream in parallel. (A link between the two may even be partially explained by the fact that large reductions in deforestation in Brazil have gone hand in hand with increasing discoveries and production of oil there.) Moreover, the deforestation problem is far from solved, for while the *rate* has been falling, we're still clearing vast tracts of carbon-rich and ecologically precious forests. In the last decade, an average of 13 million hectares of forests – an area the size of Louisiana, or half of the UK – was cleared each year.[5]

Nonetheless, the reduction in deforestation is a positive development. It suggests that humankind might finally be starting to break the back of a problem that not only contributes to climate change but is also a crucial threat to biodiversity. This progress is particularly encouraging given three significant global trends that are driving up food prices and placing extra pressure on land. The first is rising population. More than a billion additional people have been added to the global population since 1999 and there will most likely be around two billion more mouths to feed by 2050.

The second trend is meat consumption, which has been growing much faster than population growth. This has been driven almost entirely by increases in developing countries where diets are starting to become more like those of the rich world. Meat matters to forests because the world's livestock sector is hugely inefficient in its use of land and grain. It takes on average around three kilograms of grain to create one kilogram of grain-fed meat (the ratio is much higher for cows but lower for chickens) and a kilo of meat also contains fewer calories than the same weight of grain.[6] Indeed, as the diagram later this chapter shows, around a quarter of the world's food harvest – around 1,200 calories for each person on the planet – is lost through the inefficiency of current meat and dairy production.

Livestock farming doesn't have to be so inefficient. As Simon Fairlie argues in *Meat: A Benign Extravagance,* some forms are relatively efficient to produce – such as when chickens and pigs are fed on low-value crop waste. And, as mentioned earlier, there are forms of pasture-fed cattle farming, pioneered in Africa by Allan Savory, and currently undergoing large-scale adoption in Australia, that promise to be carbon-negative, locking carbon dixoide into the soil and in the process 'regrassing' land. At present, however, most global cattle production adds pressure on global forests both indirectly – by driving up food prices thereby making it more profitable to clear forests – and directly, since most previously forested land in the Amazon is eventually used for livestock pasture or growing animal feed.[7]

The third trend adding to pressure on the food and land system is an increase in the production of biofuels: petrol and diesel substitutes that can be made by processing staples such as corn, sugar and palm oil. As Paul McMahon notes in his recent book *Feeding Frenzy*, biofuels are not new. In the 1920s, as much as a quarter of American arable land was used to produce the transport fuel of the day: horse feed.[8] But in a world with a billion cars and gradually tightening carbon policies, biofuels are on the up. At the last count, around 120 billion litres of bioethanol and 30 billion litres of biodiesel were being produced each year.[9] That's almost half a litre of fuel per person per week. And although in the future biofuels could be produced from sustainably managed woodland, marginal land or even pools of algae in the ocean, the easiest option is to use food crops.

At the moment, much of the biofuel production is driven by government subsidies designed to tackle climate change – which is ironic given that many current biofuels have a bigger carbon footprint than the oil they're designed to replace. But even if these kinds of perverse subsidies were removed, biofuel production is likely to increase as our carbon-constrained world becomes ever-more thirsty for fossil fuel substitutes. The OECD-FAO *Agricultural Outlook* foresees biofuel production climbing by around 50 per cent by 2021 but the figure would surely be much higher given a massive effort to constrain global fossil fuel use.

These three trends will loom large as governments seek to continue the decline in deforestation emissions. As with fossil fuels, demand-side measures – such as efforts to ease consumption of meat, or hardwood consumption, or to reduce food waste – are necessary but insufficient. The real key is political will. Much deforestation is already illegal, so the challenge is as much about implementing existing rules as it is about creating new ones. This is tough because deforestation usually happens in remote areas, hundreds or even thousands of miles away from major cities. Moreover, the process is often incremental. Small-scale logging leads to a dirt road, which in turn leads to local settlements and small-scale clearances for farming. A bigger road may follow, which may pave the way for larger-scale farming, logging or mining. For campaigners on the ground, resisting this process can be deadly: in the last few years many local forest campaigners have been murdered in the Amazon and elsewhere. For governments, resisting it means overcoming the influence of powerful farming, timber and mining lobbies and investing in better surveillance infrastructure. But it can be done. Satellite monitoring and rapid response helicopter teams have helped slash the rate of deforestation in Brazil by around three-quarters since 2004.[10]

Success stories such as these show that, compared to giving up fossil fuels, slowing and ultimately stopping deforestation is a manageable task. Achieving it would be a huge boon both for the climate and for biodiversity – though it would also slow the increase of global food production by limiting the amount of land readily available. Combined with continued growth in population and demand for meat and biofuels, this would present the world with an increasingly stark choice in how to use the available land: to grow human food, raise livestock or produce energy crops. We'll come back to this point below.

Food's other warming impacts

If deforestation is moving in the right direction, the picture is less rosy for agriculture's other key climate impacts: emissions of methane and nitrous oxide, both of which have continued to rise steeply at the global level. In some cases there's significant potential for inexpensively reducing these emissions without cutting yields or changing what gets produced. As an example of the potential to be had from deploying simple and well established good practice, many farmers in China and elsewhere have an understandable but misplaced belief that adding more fertiliser – the chief cause of nitrous oxide emissions – always leads to a higher yield. In reality, beyond a certain point there is no benefit at all; in fact yields can go down while costs and emissions soar. Similarly, methane emissions could be significantly reduced by shifting rice production away from traditional flooded paddy fields.[11]

In some other cases, however, agricultural emissions are closely bound up with the choice of food being produced. The single largest source of methane is the flatulence of ruminant farm animals – almost entirely cows and sheep – which generate large amounts of the gas as a natural by-product of their digestion. In most cases, there's little that can be done about these emissions, but if the world could be persuaded through a mixture of education and policy to shift away from beef and lamb and towards pork or, better still, chicken, and, best of all, less meat and dairy altogether, that would make a big difference to methane emissions while also reducing pressure on forests. (If Allan Savory is correct and herd animals can help turn the world's grasslands into a giant carbon dioxide vacuum, as described earlier, this could turn the situation on its head, but there's a long way to go before that is both proved and significant in terms of the current food system.)

The chart opposite shows the carbon footprint of various meats and other foods, once estimates for methane, nitrous

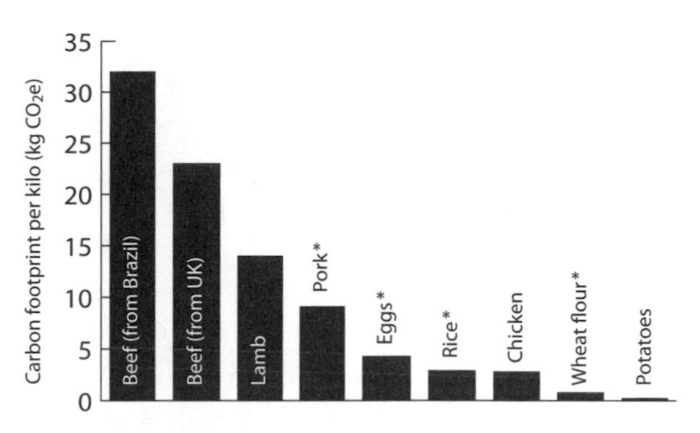

Estimates of the carbon footprint of various foods, produced for the UK market.[12] Except where indicated with an asterisk, the figures include transport to a regional distribution centre. The chart shows the warming impact over a century; beef and lamb's impacts would look even greater over a fifty-year or twenty-year timeframe.

oxide and deforestation are all taken into account. These are UK figures but the hierarchy would be similar in other countries. The numbers reveal the huge difference between the carbon emissions caused by, say, beef and chicken. They also highlight the comparative carbon-efficiency of staples such as rice, wheat and potatoes. Rice loses out to wheat mainly because of the methane emitted by flooded paddy fields. Potatoes come out well (even when you consider their relatively high water content), especially since a tenth of their energy is high-quality protein.

With better and more rigorously enforced climate regulation, some agricultural emissions would end up reflected in the cost of different foods, helping to ensure that a shift towards lower-carbon diets happens at least partly through the market. But even without such policies, a huge amount could be achieved with a combination of better technology, the spread of best practice and some shifts in diets. There's also massive scope for developing farming systems which are better at building up the carbon content of the soil.

The impact of climate change on food

Agriculture has always been humanity's greatest vulnerability to the climate and although it's impossible to predict the impacts of global warming on farming with any precision, the outlook is concerning.

Because increased carbon dioxide levels have a natural fertilising effect, it's possible that in the short term climate change will increase global agricultural yields of many staple crops. But even if that happens, there is still cause for concern, as the benefits will most likely be achieved in high-latitude areas at the expense of losses in the tropics, home to most of the world's hungry. For example, yields from African rain-fed agriculture could decline by as much as half by 2020.[13] More worryingly, at the global level, it looks like predictions of climate's impact on food may be overly optimistic. One recent study found that global warming had already reduced global maize and wheat yields by around 5 per cent.[14] This chimes with massive crop losses in 2012 caused by droughts in the US Midwest breadbasket and parts of Europe, along with a mixture of drought and downpours in the UK – all of which fit with the expected impacts of climate change. Another recent study showed that global warming will soon reduce France's maize crop unless agricultural efficiency can be significantly accelerated.[15]

Whatever happens in the short term, the IPCC predicts that total global crop yields will start to fall if the temperature climbs more than 3°C – a level that we're likely to soar past this century unless there is a massive effort to reduce emissions. Fish stocks, already under pressure, are also likely to take a hit as carbon emissions raise the acidity of the oceans and rising temperatures reduce the amount of oxygen held by the water. One recent study suggested that global warming will cause the world's fish to become 10–24 per cent smaller by 2050.[16] Once again, the impacts will be felt most strongly in the tropics.

All of this is based on scientific predictions that are full of uncertainties. Hence it's possible that the impacts could be milder than expected. But it also looks increasingly plausible that the impacts could be much worse. Recent research by James Hansen of NASA found that extreme weather events which fifty years ago would have affected, each summer, only one eight-hundredth of the world's surface now affect about around 10 per cent.[17] Those kinds of changes could hold all sort of surprises for farming.

Challenges and trade-offs

Put all this together and it's clear that food production will come under plenty of pressure this century. The world needs to accommodate billions of extra people at the same time as stopping carving farmland out of virgin forests. We need to do this while constraining the flow of fossil fuels that currently power the world's agricultural systems, while dealing with global warming's potentially serious impact on yields *and* while meeting or alleviating rising demand for meat and biofuels. To cap everything off, we need to slash methane and nitrous oxide emissions too.

It can all be done. As the diagram overleaf shows, there's currently an astonishing amount of slack and slippage on the journey from field to fork. Perhaps a third of the global harvest is currently wasted, much of it through post-harvest losses in poor countries, right where the food is most needed. Though getting the detail right on the ground is tricky, this is intrinsically a solvable problem, sometimes needing little more than the distribution of airtight plastic containers. We lose another quarter through the inefficiency of current meat production.

These are obvious areas for improvement but there's also huge potential for increasing yields in many parts of the world through all kinds of promising low-tech and high-tech techniques, some of which – such as 'regrassing' dry lands and zero-tillage arable farming – could even help get carbon from the air and into the soil. But, as with fossil fuels, the challenge with land and food is as much as anything about how we share things out – and, in particular, how much of the world's crop we feed to cars, aeroplanes and animals as opposed to people. After all, the more biofuel and conventionally produced meat we collectively consume – and the more food we throw away – the less calories there will be to feed the world's growing population. This is a serious concern because constraining fossil fuel flows will create a huge incentive to divert more crops into fuel use, and global inequality is so great that people in rich countries can comfortably absorb price rises while the world's poorest cannot.

This tension is already starting to play out. In the last few years, agricultural production has outstripped population growth but food prices have soared – in large part because a growing proportion of the world's harvest is being used to produce meat and, increasingly, biofuels. An extraordinary 40 per cent of US corn, 50 per cent of Brazilian sugar cane and 60 per cent of European vegetable oil now flows into cars not people.[18] We can get a very crude idea of the trade off between fuel and food from a basic energy analysis. A person's daily calorific requirement is about the same as the energy contained in a third of a litre of biodiesel.[19] Put this way, the 170 tonnes of fuel it takes to fill up a Boeing 747 could equate to a lot of hunger in a food- and fuel-scarce world.

The exact trade-off between fuel and food depends on the land, the crop and the efficiency of the entire supply chain. Roughly speaking, a hectare can produce the energy equivalent of five to twenty litres of mineral diesel per day, but if that energy is required in a liquid form for use as fuel, the yields are at the lower half of the range – enough to drive a small car just a hundred miles or so per day.[20] For comparison, a hectare of

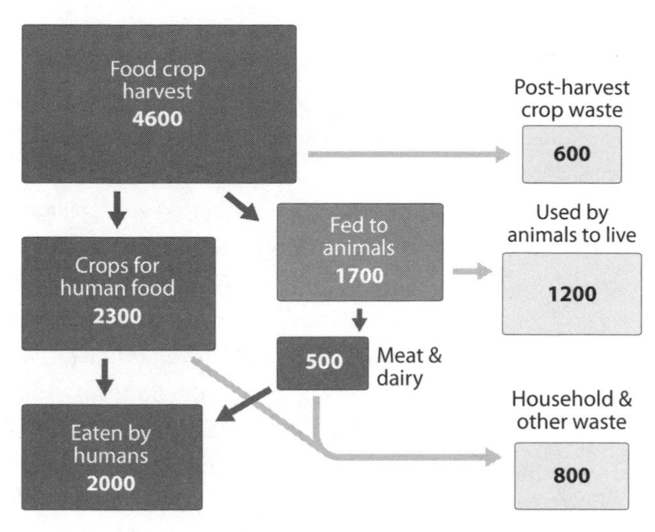

A rough map of the global food system showing calories per head. Due to losses and inefficiencies, less than half the calories produced are eventually consumed. Even after all this loss, there would still be enough to feed everyone a heathy diet if the calories were equally distributed, but at the moment there are more than a billion obese people and around a billion who go hungry. These numbers were estimated in around 2000 but probably remain broadly representative since in the interim biofuel use has risen, cancelling out production increases.[21]

fertile wheat land can provide enough calories to feed between ten and thirty people.[22]

With so many pressures on food and land, it's almost certain that prices will rise in the coming decades, just as they've risen in recent years. Higher prices tend to lead to more hunger, with the urban poor being the biggest losers. To make matters worse for those with the shortest straws, ever more land may be shifting out of their reach. Between 2000 and 2010, in excess of 200 million hectares – an area eight times the size of the UK, or nearly three times the size of Texas – was bought up in developing countries by foreign investors.[23] But even those small farmers who own their own land may feel the pinch as global warming starts to bite or as climate regulation adds to the cost of fuels and fertiliser. (According to the OECD, a 25 per cent increase in oil prices would lead to a 14 per cent rise in fertiliser prices.[24])

For all these reasons, as the climate gets hotter and emissions start to be regulated, we'll need to get better at distributing the global harvest – including carefully regulating biofuels to ensure they don't lead to more hunger. With the right policies, there's no technical barrier to having a world in which everyone is fed, emissions from farming are much lower, biodiversity is protected and some biofuel is produced to substitute for diminishing flows of crude oil. If Allan Savory's grassland management delivers as promised, perhaps we could even regenerate the world's vast arid regions and turn the global food system into a carbon sink instead of a source. But as with energy and fossil fuels, the solutions won't happen by themselves. If the world doesn't make the effort, food and land emissions could just as easily double rather than fall, while forests shrink and food prices rocket.

PART FIVE

What now?

Six key steps that will help tackle climate change

14. Waking up

Facing the facts

This book has explored some uncomfortable facts. The challenge ahead looks tough if we are to contain climate change within acceptable risks. Actions to date have barely scratched the surface. Policymakers haven't grasped what it would take to meet their own target, nor how dangerous it would be to overstep it. Citizens aren't demanding that they up their game. Our aim, in laying this all out, has been to make sense of the situation and show the need for urgent and effective action. This final section asks what that action could actually look like – and what it will take to get it started.

The following chapters look at policies, technologies and campaigns that could help. But for these to be realistic, we first need to wake up in much larger numbers to the threat. It has become a cliché to say that we could tackle climate change if we would just mobilise 'as if for war'. Yet it's true: innovations that seem decades off may prove to be just around the corner if we push for them as hard as others once pushed for radar, nuclear bombs and decoding techniques. Mass deployments could also happen fast if society sought to do so with the effort that nations have sometimes put into rolling out tanks and planes. Where the analogy breaks down is the conspicuous absence of a wartime sense of alarm.

Even many of the best-informed political and business leaders remain complacent about the risks. Fatih Birol, chief economist of the IEA, observed of the 2013 World Economic Forum that 'many people, to my surprise ... people who are very much engaged in the climate science debate from the energy and business sector, don't seem uncomfortable with the 4°C trajectory.' He added bluntly, 'This is very bad.'[1]

Those concerned about climate change need to push things forward and overcome the inertia and complacency. But how can this be done? Part of the answer is surely about leadership: people at all levels of society getting up and doing *something*. We'll discuss this more in the final chapter. Another part of the answer, we think, is for all those who care about the issue to do a better job of conveying the true nature of the situation.

That may sound obvious, but many of those who would like to see climate change dealt with have sidestepped the core issues and instead tried to sell the idea that green energy choices will save us money, solve fuel poverty and grow the economy. It was reported recently, for example, that Barack Obama made a conscious decision at the start of his first term as president to talk about climate change solely in terms of innovation, energy independence and economic progress.[2] Many companies have taken a similar line – especially in how they talk to customers – and some NGOs have even framed whole campaigns around renewable energy's potential to reduce domestic energy bills.

This approach is understandable. As well as an effort to make things more palatable it is an attempt to create a positive vision – to follow in Martin Luther King's footsteps and convey a dream rather than a nightmare. But King's audience was already fired up and demanding change – so much so in fact that in the 'I have a dream' speech he appealed to civil rights campaigners to avoid letting their frustrations spill over into violence.[3] By contrast, most people don't yet feel even slightly passionate about climate change and a vision based on the economic benefits of home insulation, smart grids and anaerobic digestion seems unlikely to change that. Anyhow, as we've shown in this book, technologies and efficiency gains in themselves won't be enough to cut emissions any time soon – especially if they're taken purely for financial gain, which makes them particularly prone to rebound effects. Moreover, the idea that rapidly constraining fossil fuel use will make us financially richer in the near term simply isn't robust; if the debate is had on those terms, the short-term economy will win and the climate will lose.

A more fundamental problem with framing most climate solutions in terms of their economic benefit may be that it reinforces unhelpful paradigms, encouraging people to think *more* about fossil fuel prices or national energy security, thereby making them more resistant to price rises caused by climate legislation. This paradox was neatly argued in *Common Cause*, an influential report written by WWF campaigner Tom Crompton.[4] There's some academic evidence to support the idea that taking apparently green steps without caring about the underlying issues can have unexpected and unhelpful side-effects. In one study, for example, groups of people were given information about car-pooling that emphasised either the financial or the environmental benefits. Those who received and read the money-saving version were found to be markedly less likely to recycle their paper forms at the end of the session, even compared to the control group who didn't see any green messages.[5]

Another common hope is that perceptions of the issue may start to shift in earnest as the climate warms – but that method of waking up is unlikely to come soon enough to get us out of trouble since further warming will continue for decades after we begin to take serious action. If we wait for the atmosphere to force us into emergency-mode, it will almost certainly be very costly and it might be too late. In any case, it seems that even extreme weather events won't necessarily be enough to trigger citizens or politicians to wake up to the issue. As we write this, Australia is being ravaged by floods only weeks after dealing with a devastating heat wave; but the front-runner in the polls for the forthcoming general election is still promising to dismantle the country's modest carbon tax system if elected.[6] So while it's important to line up thoughts and plans about how best to respond to a climate-induced alarm call, we shouldn't make that plan A.[7]

Given where we are now, it's crucial that more people hear the simple facts loud and clear: that climate change presents huge risks; that our efforts to solve it so far haven't worked; and that there's a moral imperative to constrain unabated fossil fuel use on behalf of current and especially future generations. It's often assumed that the world isn't ready for this kind of message – that it's too negative or scary or confrontational. But reality needs facing head on – and anyhow the truth may be more interesting and inspiring than the watered down version.

It's perhaps telling that for all the countless words about clean technology or personal carbon footprints, the first climate change article to go truly viral online was the Bill McKibben piece excerpted at the start of this book – a passionate, no-holds-barred take on the failure to cut emissions and the need to overcome fossil fuel interests and leave oil, coal and gas in the ground. Off the back of that article, campaign group 350.org toured the US taking a similarly hard-hitting message to university campuses. They were amazed at the response and in many cases forced to upgrade venues to accommodate the thousands of students determined to attend each show.

If this marks the start of a more truthful and upfront conversation about climate change, then it's not ahead of time. It is essential that society stops tiptoeing around the real issues. We need to insist that our media puts us properly in the picture, and that our politicians and our businesses take coherent action that matches whatever rhetoric they adopt. Rapid change in society has always required large numbers of people demanding it – and taking to the streets where necessary – on the basis of passionate concern for a big moral issue. With so much power and money bound up with fossil fuel use, climate change will surely be no different.

The right policies and technologies are important, but the ten trillion-dollar question is whether we can emerge from our collective slumber, roll up our sleeves and make it all happen.

15. Capping the carbon

How to deal with the abundance of fossil fuel

Given the feedback loops and rebound effects explored earlier, it's clear that the only reliable way to stay within a sensible global carbon budget is a worldwide agreement designed expressly for the purpose. So far, however, the idea of an all-time carbon budget hasn't seriously entered the global talks. Instead, there's just a vague hope that voluntary national pledges will one day add up to what's required. With each year that goes by, that looks less and less plausible. The negotiations are deficient in other ways too. They're attended by insufficiently senior policymakers and they're entrenched in an outdated and oversimplified view of the developed and developing world. The whole process is crying out for a reboot.

Much of the discussion of what that reboot should look like is focused on the nitty-gritty of global carbon policy – such as the relative benefits of carbon taxes versus cap-and-trade schemes. We'll come back to those questions later but for now the pros and cons of different global policy options are somewhat academic because all of them involve rapidly constraining fossil fuel use and that remains politically unrealistic in the nations that matter the most: the ones that extract, burn and sell the majority of the world's oil, coal and gas. This chapter asks what it might take, in addition to a general increase in public awareness, to change that. We focus on three key areas: minimising the influence of the fossil fuel sector on politics and public opinion in carbon-rich countries; maximising the positive global influence of nations which *are* ready to do an ambitious deal; and stemming the flow of money into fossil fuel reserves and infrastructure. We'll look at the other crucial enabler in the next chapter: a massive effort to scale up low-carbon energy technologies in parallel with constraining emissions from fossil fuels.

Another thing that civil society could reasonably and realistically demand in the short term is the attendance of national leaders at the talks. The idea of second-rank ministers dealing with the future of global energy use and the planet's atmosphere would be laughable if it wasn't so worrying. Trillion-dollar assets, global geopolitics and the safety of humanity are at stake and we need a political process commensurate with that. National leaders attending wouldn't necessarily be a game-changer but it would at least make progress possible. It was only when prime ministers and presidents briefly graced the climate talks with their presence in Copenhagen in 2009 that a temperature target was agreed and a deal started to feel vaguely within reach.[1] We need them there every time – and never to let them forget that the great political legacies of our age will be forged not on short-term domestic or economic policy, but on keeping the planet safe for future generations.

Overcoming fossil fuel interests

Considering that their products are the root cause of climate change, the fossil fuel industry has so far done remarkably well at staying out of the limelight in the climate change debate – despite the fact that many companies are actively blocking progress by prospecting for new fossil fuel reserves, building new infrastructure, buying off politicians or fomenting propaganda. We showed earlier how the global politics of climate change is shaped significantly by the size of each country's fossil fuel reserves. This is partly because fuel-rich nations have more to lose economically from a global climate deal than those which rely on imported fuels. But it's also because the bigger the fossil fuel sector in a country, the more influence it tends to wield on public policy.

The significance of this is difficult to overstate. For example, it's hard to imagine how a global deal could come about until the US agrees to take more of a leadership role. But it's difficult to see how that will happen while the American political scene is flooded with money from energy companies. Although there's much that President Obama could do unilaterally to tackle climate change (such as rule against the controversial XL Pipeline to bring crude from the Canadian tar sands to the refineries and markets of the US), his hands are tied in terms of a global deal by the recalcitrance of senators and representatives in Congress, many of whom rely on campaign donations from fossil fuel companies. We saw earlier that those heavily invested in fossil fuels are spending almost half a *billion* dollars on campaign donations and lobbying every year.

Oil, coal and gas companies certainly can't take all the blame for our failure to solve climate change. But those which lobby against carbon regulation for their own benefit, buy politicians, spread misleading information, and invest their capital in expanding their reserves rather than developing carbon capture or alternative energy need to be named and shamed – and their efforts resisted by the rest of society. That resistance can take many different forms, from old-fashioned protests, petitions and consumer boycotts to funders and companies refusing to support the politicians in their pockets. Art galleries and cultural institutions could shun their sponsorship and institutional investors could demand that the fuel companies they part-own stop blocking progress – or even divest from the fossil fuel sector altogether. Campaigners on both sides of the Atlantic are now calling on pension funds, universities, churches to do just this. Divesting may not have much direct financial impact because if someone is selling stocks or shares, then by definition someone else is buying them. But the rapidly growing campaign around divestment *is* sending a powerful message and highlighting the idea that fossil fuel production now has an unavoidable moral dimension, much like other controversial sectors such as firearms and tobacco. The idea is gaining traction, too. In December 2012 the mayor of Seattle become one of the first prominent politicians to back the campaign, writing an open letter to the city's two main pension funds encouraging them to 'refrain from future investments in fossil fuel companies and begin the process of divesting our pension portfolio from those companies'.[2]

Another possible way to challenge the legitimacy of ongoing fossil fuel extraction may be to use the courts. Until recently it's been impossible to say how climate change affects any specific weather event. But that is starting to change. The latest attribution studies can say with a reasonable degree of confidence how much more likely a flood or wildfire was made by man-made carbon emissions, and – as climate scientist Myles Allen argues – this raises at least a theoretical possibility of those who have suffered losses as a result of global warming suing the fossil fuel sector for its contribution to those losses. Realistically, the chance of anyone winning damages remains very slight, not least because no individual company is responsible for more than the tiniest proportion of the carbon that humans have added to the air. But a recognition in the courts, anywhere in the world, that each fossil fuel company could be held liable for even a tiny slice of the consequences of climate change could be both practically and symbolically important.

How some nations can take the lead – and get others to follow

If fossil fuel interests and entrenched political positions are making progress difficult in some key countries, what should the rest of the world do about it? The obvious answer is both to ramp up the political pressure – for example, by making climate change a key issue at all international forums – and to show leadership by cutting emissions rapidly even in the absence of a global deal.

Some countries are already doing just that. But as we saw earlier, all too often the benefits are being undermined by global trade in fuels and goods, or by support for fossil fuel extraction overseas. It's half-baked at best for the UK to be cutting its own emissions while doing nothing about the fact that this is outweighed by the growth in carbon-intensive goods it imports from overseas factories; or for Australia to impose a domestic carbon tax while also expanding its coal exports; or for EU member states to be claiming leadership on global climate diplomacy while using the G20 platform to call on OPEC to increase global oil output. To really show leadership, nations need to think much more about how they can reduce emissions not only within their own borders but at the global level. Partly this is a matter of smarter analysis and more joined-up policymaking but it may also include some whole new approaches.

One area that's been gaining interest recently is the idea of combining international trade and domestic climate policies. We touched earlier on one way in which this might be possible: a system of carbon border allowances to ensure that companies importing goods from overseas are subject to comparable carbon regulation to those producing goods locally. A similar policy could be used to limit exports of fossil fuels, to avoid a race to the bottom where the oil, coal or gas gets burned wherever the environmental regulation is slackest.

One major practical challenge with tying the carbon footprints of imports and exports into climate deals is that there's no robust and reliable way to measure them.[3] Reasonable approximations could be made for carbon-intensive raw materials such as steel, iron and concrete, however, and even if such a policy was limited to these it could have an impact. Alternatively, rather than using border tariffs, importer nations could set minimum standards for key materials – such as the carbon intensity of the electricity used in aluminium production – and simply ban imports that don't meet those requirements. As Columbia University's Scott Barrett has argued, the standards could start low but quickly ratchet up – and it's possible that, as with the Montreal Protocol and other successful environmental treaties, a well designed ban could effect change without ever needing to be enforced.[4]

In principle at least, tying trade into carbon policy could not just avoid 'carbon leakage' but also soften domestic resistance to ambitious national carbon goals on the ground that such targets disadvantage local industries. Just as importantly, a sensibly designed carbon-trade policy could incentivise foreign companies and governments to increase carbon regulation in their own countries in order to remain competitive in their target markets and ensure that any carbon tax revenues are collected locally rather than overseas.

Trade is a highly sensitive area, however, and the strong trend in recent decades is towards the removal rather than the addition of tariffs and regulations. Any effort to use border controls as a means of accelerating action on climate change would invite calls of protectionism and raise thorny legal questions at the World Trade Organisation.[5] The impact on poor countries could be assessed and ameliorated, but the threat of trade wars and political fall-outs would be almost inevitable. This became abundantly clear in 2012 when the EU attempted to include international air travel in its carbon cap-and-trade scheme, meaning that foreign airlines would need to buy permits to cover their emissions. In the event, airlines from the rest of the world pushed back so hard that the EU buckled under pressure and shelved the policy for at least a year. (The airlines in question received very little public criticism about this sabotage of global climate policy, a good example of where greater consumer pressure could potentially have made a difference.) The aviation example shows that any more comprehensive carbon policy involving trade would lead to even tougher political conversations. But until we're ready to empower and push our leaders to have tough conversations, the global response to climate change will most likely remain a long way from what's required.

Carbon trade policy is only one example of how nations keen to see urgent action can help reduce emissions and incentivise others to participate in a deal. There are many others, from forcing the issue onto the table at the G8 and G20 talks through to a massive increase in efforts to help poor nations achieve energy security without the need for fossil fuels. Leaders could promise to attend the talks and call on their counterparts overseas to do the same. But what if none of this works and one or more nations continue to obstruct progress while the rest of the world is ready to move forward with a deal? A conference paper written in 2001 by three economists including Nobel laureate Joseph Stiglitz considered some options, ranging from enlisting civil society in 'a massive boycott of the rogue country's products' to 'disbarring the country from participation in international activities such as the Olympic games'.[6] It notes that 'ostracism has long been a sanction employed against those who refuse to comply with international social norms.' Despite the failure to cut emissions in the intervening decade, such ideas have rarely been discussed. Perhaps it's time they were, because – however unrealistic they might seem today – they could sharpen minds. And perhaps eventually they could even be required.

Deflating the carbon bubble

Although it's received less attention than carbon pricing and global politics, another way that governments could help move the climate conversation forward is to look much more seriously at financial regulation. As we showed earlier, money is flowing into the oil, coal and gas extraction and infrastructure sectors as if global warming had never been discovered and the world's climate target was just pie in the sky. Coal companies are still floating on the world's stock markets without even mentioning in their share prospectuses how their value might be affected by global climate legislation. This is staggering – especially given that many of the world's stock markets and pension funds are already worryingly reliant on the health of fossil fuel companies. Similarly, it's unhelpful that banks and governments continue to lend money or underwrite deals for assets that could be devalued by a global climate deal, whether that's the development of unconventional fuel reserves or the rolling out of car factories, airports and other fuel-reliant infrastructure.

If the only risk of this 'carbon bubble' was the possibility that short-sighted investors might lose their money, it might not be such a big deal. But in reality blind investor assumptions that the fossil fuel economy will continue indefinitely might turn out to be self-fulfilling by making a solution more expensive and less politically feasible.

Individual banks and investors can help by thinking more carefully about the carbon implications of their investment choices, both within and outside of the fossil fuel sector – a more nuanced version of the divestment approach discussed earlier. But the key role here may be for financial regulators. To give just one example, regulators could impose a requirement on fossil fuel companies listed on local stock exchanges to publish details of how exposed their reserves are to climate regulation, including which reserves might cease to be viable at different carbon prices. This in itself wouldn't stop the fuel being burned, but it would allow investors to see what's going on – and force fossil fuel companies to be more upfront about the implications of their plans. If the oil, coal and gas sector wanted to convince investors that their reserves weren't at risk of devaluation, they'd need to be making much greater efforts to prove that carbon capture was realistic. Such regulation would also help protect pension holders and small investors from the risks of the carbon bubble inflating too much – and then bursting.

The deal itself

Progress on the three fronts discussed above, combined with much more effort on the technologies discussed in the next chapter, would make a meaningful deal much more realistic. But once we get to that point, how should it be structured? We argued in chapter twelve that the time is right for a move beyond pledges towards a rules-based system with fewer moving parts to negotiate. But which of the various options would be best?

Economists often favour the idea of a globally consistent carbon tax since it fixes the price of extracting or burning fuels, enabling businesses and governments to plan ahead and avoiding the carbon price going too high. This approach scores well in terms of an orderly and politically feasible transition in the short term. But knowing the price of carbon in advance means not knowing the actual level of emissions. And given the feedbacks in global energy use explored earlier, it seems highly possible that we'd continue to use more fossil fuels than expected at any given tax rate, wasting precious time and requiring governments to justify hiking up the tax much further than promised to stay within any target carbon budget. Even if that didn't happen, the very concept of a tax is often unpopular even at the national level, let alone the global. (A recent review of contemporary polls showed that during the 1970s oil crises Americans were more receptive to almost anything than higher gasoline taxes, even including rationing.[7])

If we're serious about staying within a given global carbon budget, the more reliable option would be to fix the level of emissions in advance, as opposed to the price. One way to do that would be the SAFE Carbon idea discussed earlier: an obligation on companies producing fossil fuels to bury an ever-rising proportion of the carbon they extract. But that would be unlikely to be accepted by poorer nations, since it fails to reflect their low level of historical responsibility for climate change. A better candidate might therefore be a global cap-and-trade scheme in which tradable permits for the production of fossil fuels are distributed to national governments according to a simple but negotiable formula such as contraction and convergence. In this model, the number of carbon permits would be fixed in advance for each year or compliance period, limiting total fossil fuel use in line with an agreed global carbon budget. Additional permits could be made available to any company demonstrably and safely putting carbon dioxide back into the ground, kick-starting a major new effort on carbon capture technology.

Putting aside the option of rationing, pinning down emissions in advance would mean leaving the price of carbon – and therefore consumer fuel prices – to the market, raising the possibility that prices could go sky high as competition for energy kicks in. If that happened, it could pose some major political and economic challenges. But on the plus side a higher-than-expected carbon price would help accelerate efforts to develop carbon capture, alternative energy and efficient technology in just the same way as the oil price spikes of the 1970s did. And if permit prices were high, then their owners would have a strong incentive to resist any watering down of the scheme, helping make it somewhat more robust.

Ultimately, though, the different ways of structuring a global carbon deal each have pros and cons. Various models – even the current approach of arbitrary national pledges – could be made to work, given sufficient political will and the backing of the world's citizens. The crucial challenge is getting to that stage.

16. Pushing the right technologies – hard

The role of carbon capture and low-carbon energy

Regulation of fossil fuels and progress with alternative energy are closely bound up. Each policy that puts pressure on carbon emissions adds to investment in and deployment of alternatives. In parallel, each development in alternative energy makes it more economically and politically plausible to impose tighter regulation on fossil fuels. Hence while we can't rely on clean technology in itself to reduce emissions – due to the balloon-squeezing effects explored earlier – we do need *much* greater efforts to develop and roll out clean energy technology and carbon capture techniques in parallel with constraining fossil fuel use.

Carbon capture

In the course of writing this book we have come to think that the most undervalued technology in terms of unlocking international progress on climate change is carbon capture – both traditional CCS for point sources such as power plants and more futuristic ambient air capture technologies for taking carbon directly out of the atmosphere. That's not to say that either family of carbon capture technology will definitely scale up to be significant in future decades. The point is simply that the more *realistic* carbon capture seems, the less a global deal will look like economic suicide for the countries with the most value tied up in fossil fuel reserves and infrastructure. And not just that: a carbon capture revolution could even help turn the world's oil and gas sector into part of the solution. With their massive scale and expertise in drilling, piping and chemical processing, no one is better placed to dominate the world of carbon sequestration than the big oil companies. A plausible global market in carbon capture could therefore give some of the world's most powerful countries and companies – and the companies with the most to lose from scaling down fossil fuel use – a powerful incentive to support climate legislation.

Support for carbon capture so far has been lacklustre, however, with governments failing to create the right incentives and in some cases scrapping or botching promising pilot schemes. Objections from some green groups that CCS is dangerous or unnecessary haven't helped; neither has resistance from local communities near trial sites. Air-capture technologies have struggled even to get to that stage. The carbon scrubber plans developed by Klaus Lackner remain on the shelf. Part of the problem, according to Lackner, is that most investors aren't interested in ventures that could take many years to turn a profit, or which rely on a carbon regulation that can't be guaranteed to come about. Governments and concerned funders need to step in and create much faster progress on both fronts. For example, governments could oblige companies extracting oil, coal or gas on their land to invest in carbon capture technologies – or even to sequester a proportion of the carbon they take out, a national-scale version of the SAFE Carbon idea.

In addition to all this, governments and companies could be working *much* harder to support and develop agricultural practices that can lock more carbon into the soil.

Renewables

Carbon capture is only part of the puzzle, of course. Renewable power sources, energy-efficient technologies and energy storage technologies are also absolutely crucial – as long as they're combined with carbon regulation to make sure they substitute for fossil fuels rather than simply adding to global energy supply. Again, there's huge scope for improvement here because although modern renewables such as wind and solar are growing super-exponentially, they remain a pin-prick in terms of total global energy supply. And presently only a handful of large countries – the most prominent being Germany – have combined aggressive ramping up of renewable power with commitments to cut carbon emissions significantly in real terms.

Economists, campaigners and policy wonks disagree on the best way to support renewables. Green groups typically argue it's necessary to subsidise their deployment directly, to build scale and make up for decades of support for fossil fuels and nuclear. Many economists prefer the idea of simply taxing carbon and letting the market decide what takes its place, to avoid governments picking winners and losers. Others argue that the real key is for governments to drive money into R&D for next-generation technologies. Given the rate of change required in the global energy system, we think all of these approaches will be required. R&D may be particularly important to ensure that, when global carbon regulation is finally agreed, we have as many tricks up our sleeve as possible. But whatever combination of policies is used in each country, it's necessary once again to think holistically about the global ripple effects.

Germany's massive deployment of solar photovoltaic panels is a good example. The roll-out – made possible with huge subsidies funded by German citizens – has often elicited criticism from commentators who see solar as an overpriced technology that's economically inefficient in cloudy countries. It's true that each kilowatt of German solar has carried a relatively high financial cost, but a conventional assessment ignores the more subtle benefits. For example, by almost single-handedly scaling up the world's solar industry, Germany has massively driven down prices for those in sunnier countries, making a global deal on emissions less economically risky for those in poor tropical regions. The policy has also stimulated huge growth in the solar industry in China, creating a lobby within that key country that may push for a global deal to help expand its markets. Furthermore, it's been shown that solar panels encourage people in the buildings where they're installed to reduce their energy use. They also have an important role as a visible symbol of change.

Nuclear options

Finally, and most contentiously, what about nuclear energy? This has become a dominant question in the last few years – so much so that those concerned about climate change have spent almost as much time debating the pros and cons of nuclear as they have in thinking about reducing global fossil fuel use, often with a good deal of mud being throw along the way. The oil, coal and gas lobbies must be delighted because in the meantime there has been far less focus on actually reducing carbon emissions – which as we've shown won't happen fast enough at the global level even if we build lots more nuclear *and* renewables. The key point about the nuclear debate, therefore, is that we shouldn't allow it to distract us from the question of how to leave fossil fuels in the ground. That said, it is important to ask whether nuclear should be part of the energy mix that replaces fossil fuels, not least because it takes many years to plan and build, so decisions made now affect what happens in the 2020s when we need emissions to be falling fast.

The main downsides of nuclear are well known: the (small) risk of large nuclear incidents; the hard-to-quantify but not trivial potential for weapons proliferation and nuclear terrorism; and the challenge of safely storing waste that will remain radioactive for centuries to come. The gravity of each of these has spawned endless debate, especially since Fukushima, which can be reasonably interpreted both as a demonstration of nuclear's risks and proof that even an extremely bad incident isn't as dangerous as is often assumed.[1]

On the other side of the coin, it's clear that in the right context nuclear can be used to generate vast quantities of low-carbon power relatively quickly. France demonstrated this conclusively by switching to an almost entirely nuclear power grid in one of the most rapid roll-outs of low-carbon energy in history. Between 1978 and 1988, total French carbon emissions fell by an average of 3.7 per cent per year – and a much bigger drop relative to what might otherwise have happened. This was largely achieved by an extraordinary program of nuclear power, which has made France one of the most carbon-efficient economies in the world. It would be neither sensible not feasible for the whole world to follow France's lead, not least due to finite uranium supplies, though new reactor designs could alleviate or remove that problem.

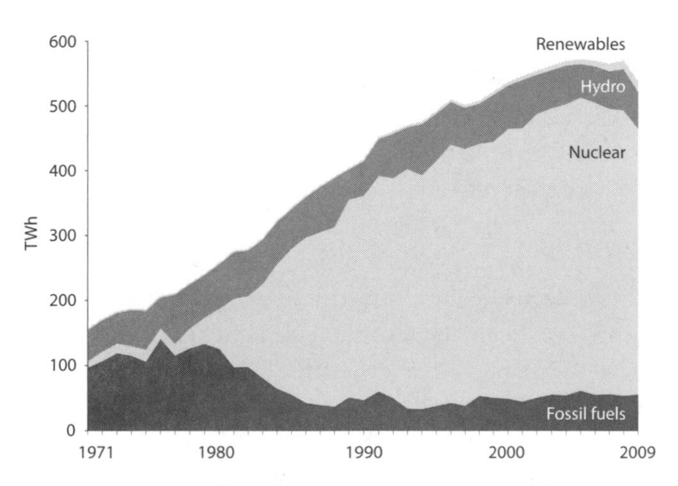

The French power grid from 1980 is a rare example of a wholesale shift in energy sources in little more than a decade. Electricity production rose steeply while fossil fuel use quickly fell. Graph based on data from the IEA.

Most people accept that nuclear can produce lots of power, however. The more important part of the debate is whether the nations would be better off investing their limited funds in renewables, efficiency, energy storage and carbon capture. Many green campaigners argue that doing so would make sense. They rightly point out that renewables are getting cheaper all the time and that some types, such as onshore wind, are already less expensive than nuclear (which, if anything, is getting pricier). They also point to studies which show that, with a global carbon deal in place, it would in principle be possible to meet anticipated energy demand without any new nuclear energy. These are legitimate arguments and yet most mainstream energy analysts believe that rejecting nuclear would make an already difficult task even tougher and more expensive – especially as nuclear is so good at providing consistent 'baseload' power. (It sometimes gets overlooked in the numbers thrown around in the media that a 1GW nuclear station will produce as much as ten times more power than 1GW of solar capacity for the simple reason that it works around the clock. To give a sense of the comparisons, the UK's biggest nuclear plant, Sizewell B can produce 8.6 trillion units of low-carbon electricity a year – the same as putting a solar roof on perhaps two to three million houses.[2])

In its 2011 *World Energy Outlook*, the International Energy Agency (IEA) concluded that achieving 50 per cent odds of avoiding a two-degree temperature rise would be a huge technical challenge even with nuclear. Without it, the IEA says, the costs would be 'substantially higher', and it's 'by no means certain' that renewables and CCS could scale up fast enough to make up the difference, 'given the scale of the imposed commercial losses and practical limitations on deploying low-carbon technologies on such a large scale and so quickly'.[3] The UK government's widely respected scientific adviser on energy, Professor David MacKay, makes a similar argument: 'If you've seriously looked at ways of making plans that add up you come to the conclusion that you need almost everything and you need it very fast – right now … I think it's unrealistic to say we could get there solely with renewables … the costs could be very high if you attempted to do that, and there's a lot of very strong resistance to many forms of renewable such as wind farms'.[4]

Unless one dismisses these perspectives as simply wrong (which seems risky at best, self-serving at worst), the question becomes whether the world is sufficiently concerned about the downsides of nuclear to add yet more cost and difficulty to implementing a global carbon deal. Our own view, given the profound threat of climate change, is that we'd be foolish to limit our options. At the very least campaigning *against* nuclear seems like an odd use of time and effort.

One area that has rightly been receiving growing attention in the last few years is the potential of next-generation nuclear technologies such as thorium liquid-salt reactors and integral fast reactors. These could offer much safer, smaller power plants that wouldn't go into meltdown in the event of a Fukushima-style power failure. Compared to conventional nuclear reactors, they can exploit incomparably more of the energy stored in their fuel – or even burn up existing high-level nuclear waste, thereby solving two problems at once. According to one recent estimate, the UK's stockpiles of plutonium and depleted uranium could in theory power the country for more than 500 years in fast reactors with fuel recycling facilities.[5]

The thorium approach was shown to work by the US government in the 1960s but then shelved, perhaps because it was less suitable than uranium-based technology for creating nuclear warheads – a disadvantage that today would be a major plus. Scientists in China, the US and elsewhere are now racing to bring the technology back. The integral fast reactor was also developed by the US government, albeit much later. The results were highly positive but the project was cancelled along with all nuclear research and development in the mid-1990s – just before concern about climate change had bubbled up. The technology has now been revived commercially and is currently being pitched to the UK government by GE Hitachi as a means of disposing of plutonium.

Whether these next-generation nuclear technologies will turn out to be commercially feasible, rapidly scalable and embraced by the public remains to be seen. They're not, as some people argue, a silver bullet for solving climate change. But they do seem sufficiently promising to be researched and trialled rather than rejected outright. Both in terms of technology and what's at stake, the context of the nuclear discussion is very different today from how it was a few decades ago and at a minimum it is important that all those with long held views consider the issue with fresh eyes.

17. Dealing with land and smoke

Farming, forests and turning down the blow torch

In parallel with limiting fossil fuel emissions, it's crucial to cut greenhouse gases and particles from other sources, including deforestation as well as methane, soot and the other fast-acting global warming 'blow torches' described in part four. A sensible global climate treaty would include provisions for tackling many of these gases and particles as a priority, but in the meantime they provide a key area for bottom-up action. The side effects are positive and the financial and political barriers relatively low. As long as such efforts don't become an excuse to ignore the fossil fuel problem, there's a huge amount of potential here – easily enough to make the difference between overstepping the two degrees threshold or not.

Methane and black carbon

A study by the United Nations Environment Programme and the World Meteorological Organisation estimated that tackling methane and black carbon alone could not only cool the climate by half a degree by 2050 but also 'avoid 2.4 million premature deaths and the loss of 1–4 per cent of the global production of maize, rice, soybean and wheat each year'.[1] The report found that substantial wins could be had from technically simple steps such as upgrading brick kilns, reducing methane leakage from gas pipelines, rolling out efficient stoves and fitting particulate filters on diesel engines. There's already progress in many of these areas; the challenge is simply speeding things up.

Forest protection

Turning to forests, we showed earlier how recent trends have been encouraging for deforestation emissions, but a much greater global effort is required to ensure they stay that way given the rising population and growing demand for animal feed and biofuels. Anything that eases those trends will help to a degree, but the real key is greater effort from governments within crucial forest countries such as Brazil, Peru and Indonesia. Just as it doesn't make sense to look at the rich world's carbon emissions without considering imports, it doesn't make sense to look at Brazilian or Indonesia emissions without considering deforestation.

As with oil, coal and gas, however, there's an important historical context to bear in mind. Rich countries cleared many of their forests centuries or even millennia ago. Although some of those forests are now expanding again, it's only fair that wealthier countries should help fund forest protection in poorer countries – and indeed to make up for any income lost by leaving the forests standing. That's already happening both directly through bilateral processes and via market mechanisms collectively referred to as REDD. So far, however, funding streams have remained very small, with almost all of the significant commitments made by a single nation: Norway.[2] Far more could be done without breaking the bank – and it would be worth it for biodiversity even if there was no climate benefit.

The Yasuní project in Ecuador provides both a glint of hope and an example of how far there is to go both on forest protection and on fossil fuel reserves. The scheme was launched after oil was discovered under the Yasuní national park, one of the most ecologically rich places on earth. Aware that oil drilling in the area would be a disaster for the forest, not to mention a major source of carbon emissions, the Ecuadorian government made the world an offer in 2010: it would leave the oil in the ground and protect the forest above if wealthier nations paid it half the value it would get for the oil over the space of thirteen years. A few years in, the project's managers are cautiously optimistic about the scheme's prospects, though only around a tenth of a target amount has so far been raised.[3] Given that the oil is sufficient to meet global demand for perhaps just ten days and that a single hectare of the forest may contain more species of trees, birds, reptiles and amphibians than the whole of North America, it's worrying that such a deal should be so hard to negotiate – and a stark reminder how tricky it will be to write-off oil that *doesn't* sit below a biodiversity hotspot.

Funding is only one part of tropical forest protection, however, and the wider world has many other roles to play. One is simply to do a better job of cutting fossil fuel use to avoid appeals for better forest protection coming across as hypocritical. Governments, companies and consumers can also avoid products associated with deforestation, such as uncertified palm oil, tropical hardwoods and imported biofuels. NGOs can contribute, too, whether that means lobbying for the land rights of indigenous forest-based people or, as the charity Cool Earth has been doing, working closely with forest communities to make sure that it's more beneficial for local people to protect rather than harm forests.[4] All of these approaches – in addition to better monitoring and enforcement efforts – will be essential to maintaining the progress seen in some regions in the last few years.

The food and land system

Solving deforestation would take care of much of agriculture's impact on the climate. But there's also a huge amount of scope for inexpensively reducing some of farming's other emissions of short-lived and long-lived greenhouse gases – and perhaps most importantly for seriously exploring carbon-negative farming. In part four we pointed out some of the biggest potential wins in cutting agriculture's footprint: better fertiliser management, smarter rice farming and, if the world's consumers could be persuaded, a move away from beef and lamb, as currently produced. However, there are many other promising approaches on the horizon, such as plants developed to be much better at fixing nitrogen from the air, reducing the need for synthetic fertilisers. In the carbon-negative arena, one particularly exciting possibility is the potential for locking vast quantities of carbon into the soil using the grazing technique pioneered by Allan Savory to mimic the action of nature's wild herds. It's remarkable how little attention this kind of approach has received so far, given the sheer scale of the claimed impact, not to mention the wide range of co-benefits.

But whatever we grow and eat, and wherever we grow it, the world will also need to work hard to improve food supply in the coming years and decades. This will be necessary not just to feed the world's growing population and counteract likely yield losses from climate change itself, but to increase the chance of getting to the point where it's possible to responsibly use productive land for producing energy or regrowing forests. Achieving this while cutting emissions would mean optimising global agriculture both for yields and ecological impact by taking the best technologies and practices from different farming systems – a concept sometimes described as 'sustainable intensification'.[5]

Getting the right skills and equipment to the world's smaller and poorer farms can be a particularly effective way of improving output and helping to boost the resilience of the most vulnerable people. On the high-tech front, key technologies include hydroponics (allowing plant growth with dramatically less water, and in doing so potentially increasing yields and making more land viable for agriculture) and GM (about which, as with the nuclear debate, we encourage renewed open mindedness from all parties). In the longer run even lab-grown meat could have potential – and there's some evidence the public would be surprisingly amenable to the idea.[6]

The other obvious way to ease pressure on land is to reduce the extraordinary amount of food that gets wasted. That includes making sure that everything that is grown is harvested and enters the food system, even if it is misshapen or would currently fall short of a picky buyer's specification. The deployment of basic storage equipment could limit losses in poorer countries and help bring food security to the places where it is most lacking. In the world's rich households, there's also a big opportunity – if the cultural barriers can be overcome – to stop so much food from being thrown away, a problem exacerbated by retailers' cautious sell-by dates and inappropriate special offers.

Greater progress on these various fronts could massively reduce the rate of global warming in the coming decades while keeping everyone well fed, protecting biodiversity and leaving as much slack as possible in the world's carbon budget for burning oil, coal and gas.

18. Making a plan B

Why geoengineering technologies are both unpalatable and worth having

The task ahead is huge and even if we get moving on it right now, there is a still a risk of failure. Even if we get a global carbon budget, it will be hard to stick within it. As Canada's last-minute exit from the Kyoto Protocol made clear, international agreements are rarely watertight. And even if we stay within a budget, we might get unlucky and find the climate warms at the upper-end of the plausible range. We already have plenty of warming locked into the system and tipping points may kick in earlier than expected. As we've seen, managing global emissions is all about juggling risk.

This is where geoengineering – technologies for tampering with the earth's climate and carbon systems, over and above simply taking carbon out of the air – could be useful to have in reserve. There are two main approaches, the first of which involves trying to use the oceans to get more carbon dioxide out of the air and store it in an inert form. Possibly the most benign option proposed along these lines is an idea called Cquestrate, which involves adding almost unimaginable quantities of lime to the sea. This would increase the amount of carbon taken out of the air while also reducing the other main side effect of our carbon emissions: ocean acidity. But while the idea looks good in principle, producing enough lime from limestone would require a vast and carbon-intensive industry in itself, so the plan would only make much sense if CCS was installed to capture those emissions. Some early estimates also suggest the cost could be higher than other means of capturing carbon from the air.[1]

An alternative approach is iron fertilisation, which involves dumping iron filings into the water to encourage plankton to 'bloom', absorbing carbon dioxide from the atmosphere before sinking to the seabed and locking the carbon away there for hundreds of years. This idea was often dismissed as unworkable until recently, but a study published in 2012 suggested that it could in fact be relatively effective.[2] But even if there were no risks of side effects, it's likely that iron fertilisation would be able to absorb little more than one gigatonne of carbon a year according to one leading researcher – only a fraction of current emissions.[3] Whether that would be a big enough win to risk tampering with the world's ocean chemistry is a matter for debate, but continued research seems sensible.

A more radical and rapid way to cool the planet would be to screen its surface from the sun. This could be done by, for example, seeding more reflective clouds over the oceans, injecting aerosols into the stratosphere, or even putting reflective particles in orbit. In principle this would work – at least for reducing the temperature – but we wouldn't know the ecological, let alone the political, consequences in advance. For all our difficulty in conceptualising climate change, only the very least imaginative among us could fail to grasp that we would be playing with fire here. As one leading scientist recently told the *New Yorker*, 'It is hyperbolic to say this, but no less true: when you start to reflect light away from the planet, you can easily imagine a chain of events that would extinguish life on earth.'[4] Even if no such risks existed, sun-screening technologies do nothing to deal with the major problem of ocean acidity, since they don't actually reduce the amount of carbon dioxide in the air.

We should be under no illusion that if we have to resort to these kinds of technologies, then humankind is in a mess. But unattractive as they are, we should be planning for radical contingencies including researching sun-screening technologies and progressing discussion of governance issues. If the climate starts running away with itself, we will regret not having every trick up our sleeve.

The second element of back-up is recognising that should the climate really go wild, there will be an uncomfortable situation to manage. Environmentally there may be a raft of consequences to try to fend off, not least a biodiversity collapse. There will be many mouths to feed amid a likely decline in global agricultural output. There is the possibility that temperature changes will bring about serious disruption to essential urban infrastructure including electricity supplies and, with them water, sewage and communications. There will be global social order to try to maintain. We can wish away this rather apocalyptic possibility, but it would probably be sensible to consider and plan for all scenarios. That includes thinking about the kind of skills and attitudes our children may need, as well as how the politics of a full scale global crisis could be handled. If the discomfort of having these conversations sharpens out motivation to make them redundant, so much the better.

19. What can *I* do?

A key role for everyone

Much of our collective failure to tackle climate change so far comes down to a lack of leadership. At all levels of society, there simply aren't enough people taking the reins and doing what they can to steer the world in the right direction. There's a vacuum to be filled here and all of us can help fill it – not just politicians and business leaders, though they're absolutely crucial, but parents, children, investors, builders, councillors, artists, lawyers. Anyone.

Leadership can take many different forms. Voters can put pressure on their representatives. Investors can rethink their portfolios and stop putting pressure on fossil fuel companies to focus on expanding their reserves. The wealthy can fund campaign groups. Families can reduce their footprints. Business leaders can stick their necks out and demand political change, or bring the climate agenda properly into their organisations so that it has real bearing on the business model. Employees and customers can demand and expect it of them. It's for each person to decide how much effort to make and which battles to pick. There are no hard and fast rules except that doing *something* is better than doing nothing – and that if it feels invigorating it's probably both more sustainable and more catchy.

Many of us feel that we're too insignificant to make a difference, but the social and political ripple effects of our efforts may be more powerful than we'd expect. After all, human society is every bit as much a complex system as the climate itself. Everyone is influenced by everyone else, and most of us are only a few degrees of separation from someone in a prominent role. Every helpful action or comment lubricates every other; every unhelpful action is a brake on progress.

This applies to words as much as it does to action. If someone challenges a nonsense fact, a half-baked policy or a greenwash strapline – in a management meeting, in a bar, on the radio, in the boardroom or in government – it becomes a fraction easier for everyone else to do the same. If a staff member at a mainstream media outlet emails her boss to say she is uncomfortable with the company's misreporting of climate change, that's worth a hundred emails from green campaigners. The coverage may start to shift just a fraction. If a pension fund manager at a mining company's AGM demands to know how the firm's plans fit with the world's stated 2°C temperature target, everyone in the room will find themselves considering the question of unburnable reserves.

The impacts don't always have to be direct. If an advertising executive declines a job to promote the Canadian tar sands, although another agency will take it up, the exchange might inspire some soul searching at the oil company and raise eyebrows in the media sector. If a student demands that a university divests from fossil fuels, although someone else will buy the shares, influential academics will be forced to think harder about the moral implications of oil, coal and gas production. Conversations will be triggered.

The same thing applies to practical carbon-cutting efforts, which can add integrity to calls for change and turn homes and companies into laboratories for finding out what works and what doesn't. If someone starts lift-sharing to cut their footprint, that will make other colleagues think – or at the very least make it clear that if they were ever to try it themselves, it wouldn't be so unusual. If an ambitious business person from an emerging economy travels to America or Europe and finds their wealthy counterparts cycling to work by choice or avoiding unnecessary business trips due to the carbon footprint, that sends a message. This is the real benefit of local actions: rebound effects might cancel out the direct carbon benefits, but the cultural ripples have the potential to multiply.

Will any of these individual efforts make a difference? It's impossible to know. But perhaps to borrow Bobby Kennedy's phrase, these 'tiny ripples of hope' will 'build a current which can sweep down the mightiest walls.' Just as each tonne of carbon we add to the air pushes the climate system closer to a physical tipping point, each positive action pushes the human system closer to a tipping point in its response. And while feedback loops are working against us in the climate, they can work for us in society. Culture change and campaigning creates political space to change laws, which can build markets, which can scale technologies, which can feed back into culture change, enabling better laws, bigger markets, and so on.

The key question of our era is which complex system will tip first, the climate or the human response. It's the ultimate high-stakes race. The climate has a big head start as it's already built-up enough momentum to make decades of further warming inevitable. It could easily tip, perhaps without us even knowing it has done so, while we're still debating the existence of a problem. But the human system has the potential to be more fleet-footed. If we can overcome the social inertia, political fear and economic vested interests, the potential is there for things to happen fast. The more of us who demand change, sound the alarm or seek to reduce fossil fuel use, the more chance we'll have of success. Deals and technologies that look many years away could happen almost overnight if enough of us decide to try and make them happen.

Alternatively, we could keep on as we are: ignoring or playing down the risks and putting responsibility for action elsewhere. But that would mean taking a monumental gamble with our children's future, and a species as intelligent as ours surely wouldn't do *that.*

Would it?

Acknowledgements

This book draws on the work of countless scientists and researchers in many different disciplines from all over the world. Our thanks go to all of them, but in particular those who have given up their time to share their expertise with us. That includes Andy Jarvis, whose paper on exponential carbon emissions was a key influence; Myles Allen, who pioneered work on cumulative emissions and has been an intellectual inspiration on topics ranging from global carbon policies to the legal implications of climate attribution; James Leaton, Christophe McGlade and Bryan Lovell for advice on fossil fuel reserves and the energy industry; Glen Peters for data on emissions embedded in global trade; Kevin Anderson for candid advice on emissions trajectories; Alice Bows and Nick Hewit for insights into the food and land system; and Michael Jacob and Alex Bowen for advice on climate economics.

Many other colleagues, acquaintances and friends have helped by talking or reading things through or chipping in ideas, either knowingly or unknowingly. These people are too numerous to list but include Adam Vaughan, Andrew Dickson, Andrew Meikle, Anthony Davis, Anthony Smith, Bryony Worthington, Chris Goodall, Damian Carrington, Daniel Vockins, David Parkinson, Edwin Booth, Fred Pearce, Ian Katz, Jeremy Leggett, John Urry, Judi Marshal, Kim Studwick, Leo Hickman (who came up with the 'balloon squeezing' metaphor), Mark Lynas, Phil Latham, Roland Chambers, Tom Clark, Tristan Smith and Warren Hatter and the staff and students of the UCL Energy Institute.

Others gave practical assistance. Thanks to Rosemary Grant-Muller for pulling together essential data sets; James McKinstry-West for straightening out endless references; Rachel Garside and Chris Kempster for miscellaneous bits of analysis; and the staff at Quaker Café by London's Euston station for accommodating us for countless hours of thinking and writing. On the publishing side, we're grateful to everyone at Profile, but especially our editor Mark Ellingham for signing up the book and helping improve it immeasurably. Very special thanks also go to Bill McKibben, both for providing the foreword and for doing such a remarkable job of keeping up the pressure on climate change in the US and wider world.

Finally, thanks to those people who put up with the side effects of the book's evolution. That includes Robin Houston who tolerated seemingly endless interference with the development of his and Duncan's business, Kiln, and everyone at Small World Consulting for putting up with Mike's distractions and oddball questions. Above all, though, thanks to our families. This book created more than its fair share of tiredness, stress and absence over many months and at all times of night and day. We wouldn't have been able to complete it without the patience, support and encouragement of Bill, Elizabeth, Eva, Liz and Rosie.

Notes and references

To aid readability and usability, these endnotes include forwarding web addresses. The underlying link will become visible in your browser after the forwarding address is entered. These notes are also reproduced online at burningquestion.info/notes.

Foreword

1. B. McKibben, 'Global Warming's Terrifying New Math' (*Rolling Stone*, 2012, bitly.com/new-math).

Chapter 1: The curve

1. For a readable summary of fire's role in human brain evolution, see R. Wrangham, *Catching Fire* (Profile Books, 2009). Diet also contributed to brain development much earlier – e.g. when primates developed colour vision, helping them spot ripe fruits amid green foliage, as described in W. R. Leonard *et al.*, 'Effects of brain evolution on human nutrition and metabolism' (*Annual Review of Nutrition*, 2007, bitly.com/brain-energy).
2. Marco Polo, *The Travels of Marco Polo*, Volume One, Chapter XXX. Available online at: bitly.com/mp-travels.
3. W. S. Jevons, *The Coal Question* (1865). The full quote runs: 'Coal in truth stands not beside but entirely above all other commodities. It is the material energy of the country – the universal aid – the factor in everything we do. With coal almost any feat is possible or easy; without it we are thrown back into the laborious poverty of early times. With such facts familiarly before us, it can be no matter of surprise that year by year we make larger draughts upon a material of such myriad qualities – of such miraculous powers.'

4. 'More than 500 Exajoules' from GEA, *2012: Global Energy Assessment* (International Institute for Applied Systems Analysis, 2012, bitly.com/GEA-2012). Estimates for the amount of energy that a working human can provide typically clock in at around 100W – enough to power a single old-fashioned light bulb. That equates to 360,000 joules per hour. Working seven hours a day, five days a week, that would give around 650 megajoules per 'energy slave' per year. Average per capita energy use is currently around 71 gigajoules (500 exajoules / 7 billion), equivalent to around 110 such energy slaves.

5. And the rate of increase of that rate of increase is proportional to the rate of increase. And so on forever.

6. Andrew Jarvis and colleagues at Lancaster University pointed out the fit of manmade emissions to an exponential curve with, statistically speaking, astonishingly little deviation. The same applies to energy use. Here is the maths from their paper (which unlike the chart in chapter one excludes cement production):

   $U = e^{\alpha(t-t_1)}$

   Where U is annual global anthropogenic CO_2 emissions, from energy and land use change combined (in gigatonnes per year), $\alpha = 0.0179 \pm 0.0006$ per year (the very small error margin from exponential shows the strength of fit to an exponential curve) and $t_1 = 1883\text{AD} \pm 1.7$ years

   And for energy use, where E = annual energy use in joules $\times 10^{15}$

   $E = e^{\mu(t-t_1)}$

   Where $\mu = 0.0238 \pm 0.0008$ and $t_1 = 1775\,\text{AD} \pm 3.5$ years

   Note that μ is greater than *α*. In other words carbon emissions have been less aggressively exponential than energy use, showing that we have been steadily decreasing the carbon intensity of energy over time, as described in chapter seven.

   Source: A. Jarvis *et al.*, 'Climate–Society Feedbacks and the avoidance of Dangerous Climate Change' (*Nature Climate*

Change, 2012, bitly.com/carbon-curve). That paper in turn draws on the most definitive datasets for human carbon emissions: T. A. Boden *et al.*, 'Global, Regional, and National Fossil-Fuel: CO_2 Emissions' and R. A. Houghton, 'Carbon Flux to the Atmosphere from Land-Use Changes: 1850–2005' (both Carbon Dioxide Information Analysis Center, 2010).

7. 'Trends in Global CO_2 Emissions; 2012 Report' (PBL Netherlands Environmental Assessment Agency, 2012, bitly.com/neaa-2012)

8. After 600 years, our emissions would be around 30 GT (today's emissions) $\times e^{(0.0179 \times 600)} = 1.38 \times 10^{15}$ tonnes. Of that, 32/44 would be oxygen $= 1.01 \times 10^{15}$ tonnes. That's more than there is oxygen in the atmosphere, which would be 20 per cent of the total weight (5.3×10^{18} kg, according to bitly.com/atmos-mass) $= 1.06 \times 10^{15}$ tonnes.

Chapter 2: Heading for trouble fast

1. For example, see 'G8+5 Academies' Joint Statement: Climate Change and the Transformation of Energy Technologies for a Low Carbon Future' (2009, bitly.com/academy-statement). It says that it is 'essential that world leaders agree on the emission reductions needed to combat negative consequences of anthropogenic climate change'. The statement is signed by the heads of Academia Brasileira de Ciéncias, Brazil; Royal Society of Canada, Canada; Chinese Academy of Sciences, China; Académie des Sciences, France; Deutsche Akademie der Naturforscher Leopoldina, Germany; Indian National Science Academy, India; Accademia Nazionale dei Lincei, Italy; Science Council of Japan, Japan; Academia Mexicana de Ciencias, Mexico; Russian Academy of Sciences, Russia; Academy of Science of South Africa, South Africa; Royal Society, United Kingdom; and National Academy of Sciences, United States of America.

2. 'January 2013 Global Temperature Update' (NOAA National Climatic Data Center, 2013, bitly.com/335th)

3. For an excellent layman-friendly summary of the science, see R. Henson, *The Rough Guide to Climate Change* (Rough Guides/Penguin, 2011). To go straight to the science sources, read the most recent IPCC report (available at www.ipcc.ch and surprisingly layman-friendly). For additional corroboration, see the Berkeley Earth project (berkeleyearth.org), run by physicist and ex-climate sceptic Professor Richard Muller, whose team examined 250 years of temperature records from scratch and concluded that the mainstream view of warming was correct and that 'humans are almost entirely the cause'.

4. So speculated Martin Rees, former president of the UK's Royal Society, in a speech at the opening of the Gordon Manley building at Lancaster University in 2006.

5. The IPCC latest report points to a 0.74°C temperature rise, which equates to an increase of 1.33°F. Source: *Climate Change 2007: Synthesis Report* (IPCC, 2007, bitly.com/IPCC-2007)

6. For example: R. Tol, 'The Economic Effects of Climate Change' (*Journal of Economic Perspectives*, 2009, currently available at: bitly.com/tol-2009).

7. IPCC *Fourth Assessment Report, Working Group II*: Impacts, Adaptation and Vulnerability, Table 19.1 (2007, bitly.com/ipcc-agriculture).

8. P. C. Tzedakis *et al.*, 'Determining the Natural Length of the Current Interglacial' (*Nature Geoscience*, 2012, bitly.com/tzedakis-2012).

9. James Hansen and colleagues at NASA found that events extreme enough to affect only 0.13 per cent of the earth's surface each summer in the period 1951 to 1980 (at the time these were three standard deviations away from average weather) are now affecting a massive 10 per cent of the earth's surface each year. Hence they were able to conclude: 'We can state, with a high degree of confidence, that extreme anomalies such as those in Texas and Oklahoma in 2011 and Moscow in 2010 were a consequence of global warming because their likelihood in the absence of global

warming was exceedingly small.' Source: Hansen *et al.*, 'Public Perception of Climate Change and the New Climate Dice' (*PNAS*, 2012, bitly.com/hansen-2012).

Other important studies looking at attribution of extreme weather events to climate change include: P. Pall *et al.*, 'Anthropogenic Greenhouse Gas Contribution to Flood Risk in England and Wales in Autumn 2000' (*Nature*, 2011, bitly.com/pardeep-2011).

10. A readable summary why this is with references to key scientific papers is available from the NASA Earth Observatory at bitly.com/climate-lag.

11. A. Davis *et al.*, 'The Impact of Climate Change on Indigenous Arabica Coffee (Coffea arabica): Predicting Future Trends and Identifying Priorities' (*PLoS ONE*, 2012, bitly.com/climate-coffee).

12. For a page-turning potted history of climate science, along with a candid description of the risks of abrupt change and a persuasive call for carbon scrubbing, read *Fixing Climate: The Story of Climate Science and How to Stop Global Warming* by Robert Kunzig and Wallace Broecker (Profile Books, 2008).

13. The official agreement also nods to 1.5°C as a way of acknowledging that 2°C might submerge a number of low-lying island nations, but virtually no policymakers are pushing for such a target to be met.

14. J. B. Smith *et al.*, 'Assessing Dangerous Climate Change through an Update of the Intergovernmental Panel on Climate Change (IPCC) "Reasons for Concern"' and M. Mann, 'Defining Dangerous Anthropogenic Interference' (both *PNAS*, 2009).

15. J. Hansen and M. Sato, 'Paleoclimate implications for Human-made Climate Change'. In: *Climate Change: Inferences from Paleoclimate and Regional Aspects* (Springer, 2012, currently available at: bitly.com/hansen-dangers).

16. See: J. Fasullu and K. Trenberth, 'A Less Cloudy Future: The Role of Subtropical Subsidence in Climate Sensitivity' (*Science*, 2012,

bitly.com/less-cloudy). The US National Center for Atmospheric Research summed up the research as follows: 'Climate model projections showing a greater rise in global temperature are likely to prove more accurate than those showing a lesser rise' (source: bitly.com/ncar-models).

17. The IPCC estimates that the Last Glacial Maximum was 3–5°C cooler than the present, which roughly equates to 2.25–4.25°C cooler than the preindustrial temperature. Source: *Climate Change 2007: Working Group I: The Physical Science Basis* (IPCC, 2007, bitly.com/ipcc-ice).

18. 'Extreme temperatures' (Met Office, 2011, bitly.com/prediction-sources). Also see the Met Office's visualisation of impacts at 4°C (bitly.com/met-4°c).

19. K. Anderson, 'Climate Change: Going Beyond Dangerous' – a presentation with slides available at: bitly.com/kevin-anderson.

20. She said: 'The current state of affairs is unacceptable precisely because we have a responsibility and a golden opportunity to act. Energy-related CO_2 emissions are at historic highs, and under current policies, we estimate that energy use and CO_2 emissions would increase by a third by 2020, and almost double by 2050. This would be likely to send global temperatures at least 6C higher within this century.' Source: F. Harvey and D. Carrington, 'Governments Failing to Avert Catastrophic Climate Change, IEA Warns' (*The Guardian*, 2012, bitly.com/iea-6c).

21. Foreword from 'Turn Down the Heat: Why a 4°C Warmer World Must be Avoided' (World Bank, 2012, bitly.com/turn-down-heat).

Chapter 3: The trillion-tonne limit

1. M. Allen *et al.*, 'Warming Caused by Cumulative Carbon Emissions Towards the Trillionth Tonne' (*Nature*, 2009, bitly.com/allen-2009). Other studies on the same topic have come up with similar estimates. These include M. Meinshausen *et al.*, 'Greenhouse-gas emission targets for limiting global warming to

2°C' (*Nature*, 2009, bit.ly/meinshausen). In this book we stick with the trillionth tonne paper for simplicity, though it's important to note that some other papers have suggested a slightly lower temperature response to any cumulative emissions budget. See for example: H. Damon Matthews *et al.*, 'The proportionality of global warming to cumulative carbon emissions' (*Nature*, 2009, http://bit.ly/zSTmZa). Recent developments in estimates for climate sensitivity, informed by the rate of temperature change in the last decade, may support these more optimistic assessments, which would – encouragingly – suggest slightly larger budgets for any given odds of hitting 2°C.

2. Confusingly, 'carbon' is often used as shorthand for carbon dioxide but when it comes to weights the two are quite different. Each tonne of carbon atoms in oil, coal or gas becomes 3.7 tonnes of carbon dioxide molecules. This is simply because the CO_2 contains the weight of two oxygen atoms as well as one carbon atom.

Chapter 4: Too much fuel in the ground

1. Exact terminology and definitions for different types of resource and reserve vary widely between country and company. Sometimes the percentages describe the chance of being produced rather than the chance of being exceeded. Proven are often described as '1P' or 'P90'; proven plus probable reserves as '2P' or 'P50'; and proven plus probable plus possible reserves as '3P' or 'P10'. For an excellent summary of the ambiguities and terminology, see C. McGlade, 'A Review of the Uncertainties in Estimates of Global Oil Resources' (Energy, 2012, bitly.com/reserves-resources).
2. 'BP Statistical Review of World Energy' 2011 and 2012 editions. Available at: bitly.com/bp-sr.
3. Shell was subject to a huge amount of criticism in 2004 after it turned out to have exaggerated its proven reserves by 20 per cent.

For a contemporary report on the fall-out see T. Messenger, 'Oil Giant Shell's Investors Shocked' (BBC, 2004, bitly.com/bbc-shell).

4. A US government assessment of 32 countries claimed that those nations had 169 trillion cubic metres of technically recoverable shale gas – around the same as the world's economically recoverable reserves of conventional natural gas. But official figures for the US were cut almost in half in early 2012, while Cuadrilla a company drilling in the UK recently announced its Lancashire site contains 5 trillion cubic metres – ten times more than the US estimate for the whole UK. Similarly, China's own survey put its reserves nearly twice as high as the figure given in the US survey. For links and more background see D. Clark, 'Q&A: Shale Gas and Fracking' (Guardian.co.uk, 2012, bitly.com/shale-gas-guide).

5. The chart is based on data from the IEA *World Energy Outlook 2012*. The figures roughly match those from other sources such as BP's 'Statistical Review of Energy 2012'.

6. This issue was first explored in depth in: P. Kharecha and J. Hansen, 'Implications of "Peak Oil" for atmospheric CO_2 and Climate' (*Global Biogeochemical Cycles*, 2008, bitly.com/hansen-peakoil).

7. Data supplied directly to the authors by Christophe McGlade at the UCL Energy Institute. They are being prepared for publication and aren't available publicly at the time of writing.

8. IEA, *World Energy Outlook 2012*, Table 2.4: Fossil-fuel reserves and resources by region and type, end-201.

9. For the sake of simplicity, this stack ignores certain emissions such as the 'fugitive' (leaked) emissions of methane from shale gas extraction.

10. For a good summary of the 'peakist' view, see Jeremy Leggett's *Half Gone* (Portobello Books, 2006).

11. IEA, *World Energy Outlook 2012*, Table 3.15: World oil supply by type in the New Policies Scenario. The New Policies Scenario is

the IEA's 'central scenario', which takes into account existing and announced government commitments and plans.

Chapter 5: No deal on the horizon

1. For example, one survey by academics from the School of Psychology at Cardiff University found that 'most respondents ... regard national governments (32 per cent) and the international community (30 per cent) as being mainly responsible for taking action'. Source: A. Spence *et al.*, 'Public Perceptions of Climate Change and Energy Futures in Britain' (working paper, 2012, bitly.com/spence-2010).
2. The United Nations Framework Convention on Climate Change, Article 2, Objective: 'The ultimate objective of this Convention and any related legal instruments that the Conference of the Parties may adopt is to achieve, in accordance with the relevant provisions of the Convention, stabilisation of greenhouse gas concentrations in the atmosphere at a level that would prevent dangerous anthropogenic interference with the climate system. Such a level should be achieved within a time-frame sufficient to allow ecosystems to adapt naturally to climate change, to ensure that food production is not threatened and to enable economic development to proceed in a sustainable manner.' Available at: bitly.com/unfccc-objective.
3. The resolution stated that 'the exemption for Developing Country Parties is inconsistent with the need for global action on climate change and is environmentally flawed'. The Byrd-Hagel Resolution is available online at: bitly.com/kyoto-senate.
4. Government announcements submitted under the Copenhagen Accord and acknowledged under the Cancun Agreements. Source: climateactiontracker.org.
5. The 'green paradox' concept was popularised by German environmental economist Hans-Werner Sinn. The existence and scale of the effect is the subject of debate.

6. Full details of the UNFCCC Durban Agreements are available at bitly.com/durban-agreements.

7. Map generated by D. Clark and R. Houston using data from the Climate Analysis Indicators Tool (CAIT) Version 9.0 (WRI, 2012, cait.wri.org). For an interactive version of the map, see carbonmap.org.

8. In 1997, the consumption footprint of the developed world accounted for 3831 MT, compared to 2612 MT for the developing world. The lines crossed in 2009. Source: Peters *et al.*, 'Rapid Growth in CO_2 emissions after the 2008–2009 Global Financial Crisis' (*Nature Climate Change,* 2012, bitly.com/peters-2012). Written up by Duncan for *The Guardian* at: bitly.com/poor-overtake.

Chapter 6: Rebounds and ripples

1. The Model-T got 13–21 mpg and had a top speed of 45 mph according to Ford (source: bitly.com/model-t-facts), while Fiat quote 69 mpg for their 500C Twin Air and economy is rising all the time. Some prototype cars such as VW's L1 Hybrid can get as much as 180 mpg.

2. W. S. Jevons, *The Coal Question* (1865).

3. For a good technical overview of rebound effects, see 'The Rebound Effect: An Assessment of the Evidence for Economy-wide Energy Savings from Improved Energy Efficiency' (UK Energy Research Centre, 2007, bitly.com/rebound-effect). For a highly readable summary of many elements of the debate, see David Owen's book, *The Conundrum* (Riverhead, 2012). Also worth reading are the numerous online debates between Owen and efficiency advocate Amory Lovins of the Rocky Mountain Institute, some which are linked off from: bitly.com/jevons-paradox.

4. 'Energy Consumption in the UK' (Department for Trade and Industry, undated, bitly.com/uk-energy). In the particular case

of lighting it may well be that the total energy consumption has subsequently fallen as highly efficient bulbs achieve market dominance, but even when that happens it will be clear that increased consumption has offset many of the benefits.

5. This idea of more efficient and comfortable travel as an 'amplifier' of overall energy use is explored in depth in D. Owen, *The Conundrum* (Riverhead, 2012).

6. O. Milman, 'Australian "Mega Mine" Plan Threatens Global Emissions Target' (*The Guardian*, 2012, bitly.com/megamine).

7. The plans have caused a stir in Washington State, as reported in: 'Fights Brewing over Massive Coal-Export Plans for the Northwest' (*Seattle Times*, 2012, bitly.com/coal-export).

8. 'Oil and Gas: Petroleum Licensing Guidance' (Department of Energy and Climate Change, undated, bitly.com/uk-oil).

9. The Technology Strategy Board recently announced a million pounds for SMEs that could 'stimulate innovation and accelerate the development and deployment of new technologies likely to enhance production and asset reliability within the oil and gas sector'. Source: 'Developing Innovation in the Oil and Gas Sector' (bitly.com/tsb-oil). This grant is just the latest example of a subsidy to fuel extraction. Much larger tax breaks and subsidies pervade the developed and developing world. For more information see: D. Clark, 'Fossil Fuel Subsidies and Tax Breaks Are Still Rising' (*The Guardian*, 2013, bitly.com/fuel-subsidies).

10. G. Monbiot, 'Scottish Climate Policy Is Hypocritical, Contradictory and Counter-productive' (*The Guardian*, 2009, bitly.com/monbiot-fuels).

11. Peters *et al.*, 'Growth in Emission Transfers via International Trade from 1990 to 2008' (*PNAS*, 2011, bitly.com/peters-2011).

12. 'Government Should Be Open about Outsourced Emissions' (Parliament.uk, 2012, bitly.com/outsourced-emissions).

13. Expenditure on electricity in China causes about three and half times as many emissions as the same electricity spend would have in the UK. The main reason for this is that coal is both cheaper than the UK's energy mix and more carbon intensive per kilowatt hour. The financial incentive for managing carbon within Chinese industry is therefore dramatically lower than it is in the UK, almost certainly leading to energy inefficiencies on top of the carbon intensity of the energy itself.

14. D. Guan *et al.*, 'The Gigatonne Gap in China's Carbon Dioxide Inventories' (*Nature Climate Change*, 2012, bitly.com/guan-summary).

15. House of Commons Energy and Climate Change Committee, 'Consumption Based Emissions Reporting' (House of Commons, 2012. bitly.com/uk-outsourced).

16. In itself, a switch to a global carbon framework based on consumption-based reporting wouldn't necessarily reduce emissions, since it would encourage countries to allow more heavy industries as long as they were exporting the results. This point is made by M. Jakob and R. Marschinski, 'Interpreting Trade-related CO_2 Emission Transfers' (*Nature Climate Change*, 2012, bitly.com/traded-co2). Hence it is not a panacea. But consumption-based accounting *is* necessary for seeing which countries are or aren't reducing their footprint.

17. Advocates of this approach include economist Dieter Helm, author of *The Carbon Crunch* (Yale University Press, 2012).

18. Admittedly this is an oversimplification – not least because since the financial slump emissions have actually been running below the cap. But the broad point stands: when there's a cap in place, then the action taken within that cap doesn't usually affect the total level of emissions. Whether the ETS has made much difference to emissions is a matter for debate. It has suffered from various design flaws, such as over-allocation of permits, and the result has been a carbon price too low to drive significant levels of change. On the other hand, the ETS has at least shown that a

large-area carbon trading scheme is possible. For a summary of the issues, see: 'What is the Emissions Trading Scheme and Does it Work?' (Sandbag and *The Guardian*, 2011, bitly.com/ets-carbon).

Chapter 7: People, money and technology

1. The Kaya Identity was developed by and takes its name from Yoichi Kaya who used it in papers such as: 'Impact of Carbon Dioxide Emission Control on GNP Growth: Interpretation of Proposed Scenarios' (IPCC Energy and Industry Subgroup, 1990) and Y. Kaya and K. Yokobori, '*Environment, Energy, and Economy: Strategies for Sustainability*' (United Nations University Press, 1997). The Kaya Identity is only the latest of a number of similar formulas to describe human environmental impact, the best known other one being 'I = PAT', developed by Paul Elrich and others, which states that Impact = Population × Affluence × Technology.

2. Data combined from various sources, referenced at burningquestion.info/data.

3. The World Bank data website puts the global average fertility rate at 4.92 in 1960 and 2.45 in 2010. Figures available at bitly.com/fertility-rate.

4. All figures based on mean growth rates in the period 2005–2010, calculated from World Bank data available at: bitly.com/population-data.

5. Extrapolated from the international carbon footprint database created for the paper Peters *et al.*, 'Growth in Emission Transfers via International Trade from 1990 to 2008' (*PNAS*, 2011, bitly.com/peters-2011). The data themselves are available at bitly.com/peters-data.

6. The relationship between emissions and population growth isn't always so clear cut. There are anomalies such as the Middle East, which – with its fossil fuel riches and low levels of female empowerment – has relatively high emissions *and* a high birth

rate. Moreover, populations are still growing, albeit more slowly, in many countries where carbon footprints are high, such as the US and UK. Broadly speaking, though, the trend is clear: most of the population growth is in poor countries while most of the fossil fuel use is elsewhere. For a well researched and written survey of world population trends and the role of female education and empowerment in reducing average fertility levels, see Fred Pearce's *Peoplequake: Mass Migration, Ageing Nations and the Coming Population Crash* (Eden Project Books, 2011).

7. D. Satterthwaite, 'The Implications of Population Growth and Urbanization for Climate Change' (International Institute for Environment and Development, 2009, bitly.com/satterthwaite-2009).

8. One recent academic paper, based on a model sophisticated enough to consider factors such as the economic impact of urbanisation and ageing, went so far as to suggest that limiting population growth could provide 16–29 per cent of the emissions reductions needed by 2050. O'Neill *et al.*, 'Global Demographic Trends and Future Carbon Emissions' (*PNAS*, 2010, bitly.com/oneill-2010).

9. 'World Population Prospects, 2010 Revision' (United Nations, 2010, bitly.com/un-population).

10. F. Pearce, 'Dubious Assumptions Prime Population Bomb' (*Nature*, 2011, bitly.com/pearce-nature).

11. 'Rapid Growth in Less Developed Regions' (United Nations Population Fund, undated, bitly.com/pop-trends).

12. 'Roughly' because there are slight differences between GDP per person and income per person. More accurately, GDP measures the total amount of goods and services produced and sold by all the companies and people within a country. Also commonly used is GDP's near identical twin – GNP or gross nation product – which adds up the goods and services produced by the citizens and companies of a country, wherever they are in the world. Both are incomplete measures of productive activity as they miss

out services that don't involve any payment, such as caring for relatives and volunteering.

13. 'Reasonably steadily'. Looking at an rolling trend line you could argue that there is a wave pattern corresponding with economic cycles.

14. To give just a few examples, Joseph Schumpeter believed innovation was the key; Adam Smith famously pointed to labour specialisation and improved business processes in his example of a pin factory; Tim Jackson and many environmentalists argue that cultural values are an additional important factor. Energy's role in particular has been debated. Mainstream economics has traditionally assigned it a relatively small role in driving growth on the grounds that fuel and power sales represent a small proportion of total GDP. More recently, however, ecological economists have argued that energy has played a far bigger role in driving growth than neoliberal economics suggests. This, they claim, helps explain why a small decline in the availability of energy – such as in the oil crises of the 1970s – can lead to major recessions. For a readable summary of this debate and links to various academics papers see A. Fanning, 'Economics, Growth and Energy in the Green Economy' (*Human Dimensions*, 2012. bitly.com/energy-growth).

15. R. York, 'Asymmetric Effects of Economic Growth and Decline on CO_2 Emissions' (*Nature Climate Change*, 2012, bitly.com/york-2012).

16. For an engaging summary of how rising incomes can help slow population growth, see Hans Rosling's customarily engaging TED talk 'Global Population Growth, Box by Box'. Available at: bitly.com/box-by-box.

17. In *Technological Trajectories and the Human Environment* (National Academy of Engineering, 1997) the authors estimate that the longer-term trend 'appears to have averaged about 1 percent per year since the mid-nineteenth century and about 2

percent per year in some countries since the mid-1970s'. Available online at: bitly.com/tech-trajectories.

18. Overall, low-carbon energy sources currently account for around 19 per cent of global energy consumption at the time of writing, although most of that comes from traditional wood, dung and hydro – not all of which are necessarily sustainable, let alone rapidly growing. For a recent summary of the data, see: 'Renewables 2011 Global Status Report' (REN21, 2011, bitly.com/gsr-2011).

19. Graph adapted from the IEA's *World Energy Outlook 2012*, Figure 5.2: Incremental world primary energy demand by fuel, 2001–2011. Note that renewables would look better from the perspective of consumer supply because much of the energy in coal and gas is wasted as heat in power stations. But from the climate's perspective, it's only the fossil fuel use that matters.

20. Figures are from BP's 'Statistical Review of World Energy 2012' (bp.com/statisticalreview) and are based on 'gross generation from renewable sources including wind, geothermal, solar, biomass and waste'. Hydro is not included.

21. Chart courtesy of the International Institute for Applied Systems Analysis. Source: A. Grubler *et al.*, *Global Energy Assessment: Toward a Sustainable Future* (IIASA/Cambridge University Press, 2012, globalenergyassessment.org).

22. For an engaging summary of the surging demand for air conditioning in India, see E. Rosenthal and A. W. Lehren, 'Relief in Every Window, but Global Worry Too' (*New York Times*, 2012, bitly.com/aircon-demand).

23. Conversation between Duncan Clark and fast-breeder nuclear advocate Tom Blees. Of course, if fast nuclear reactors became low-cost enough to serve this role, then arguably policymakers would find it much easier to regulate carbon and the world could use nuclear for powering and heating homes and factories directly, enabling the gas to stay in the ground. But that won't

necessarily be easy given the existing fossil fuel infrastructure and the difficulty of installing nuclear power in many countries.

24. The marginal costs and impacts of a commodity are the costs and impacts of increasing supply by one unit. This can be significantly higher or lower than the average cost or impact. In the case of coal power, the marginal cost of continuing to use an existing plant is lower than the total average cost of a unit of electricity, because money has already been sunk into the plant's construction. Marginal demand also affects all kinds of other environmental assessments. For example, even though most countries use a mix of renewables, nuclear and fossil fuels, the marginal demand is almost entirely supplied by fossil fuels, because any nuclear and renewable capacity available will usually be working flat out, given that they have no fuel costs. Hence although the average carbon footprint of a unit of electricity might be, say, 500 grams, the actual footprint of *adding* a unit to demand (or the savings of reducing demand by a unit) may be much higher – say 900 grams.

25. 'European Integrated Oils' (Deutsche Bank, 2009, available at the time of writing at bitly.com/oil-costs).

26. IEA, *World Energy Outlook 2012*, Table 2.1, World primary energy demand and energy-related CO_2 emissions by scenario.

27. Mark Scott, 'The Big New Push to Export America's Gas Bounty' (*New York Times*, 2012, bitly.com/gasexports). As one energy executive quoted in the article puts it: 'There's so much potential for the U.S. to take advantage of high prices in global markets … We've got to capitalize while we can.'

28. Different studies and reports have reached different conclusions about the amount of 'fugitive' methane emitted during fracking. One of the more pessimistic studies is R. Howarth *et al.*, 'Methane and the Greenhouse-gas Footprint of Natural Gas from Shale Formations' (*Climatic Change*, 2011, bitly.com/methane-leaks). Opinions also vary about the ease with which these emissions might be reduced. For a broader background to shale gas issues

see, D. Clark, 'Q&A: Shale Gas and Fracking' (*The Guardian*, 2012, bitly.com/shale-gas-guide).

29. The Organisation of the Petroleum Exporting Countries (www.opec.org) is a cartel of twelve nations that collectively account for a large proportion of the world's oil production. OPEC limits total exports by agreeing production quotas. This ensures that oil prices are higher than they would otherwise be as well as increasing the geopolitical influence of the nations themselves. The Organisation's principal aim is 'the coordination and unification of the petroleum policies of Member Countries and the determination of the best means for safeguarding their interests, individually and collectively'. OPEC's influence on the global oil market is smaller than it once was, due to more oil coming on stream in other countries, though its production quotas remain a significant factor in determining global energy prices.

Chapter 8: The write-off

1. L. Mageri, 'Oil: The Next Revolution', Discussion Paper 2012–10 (Belfer Center for Science and International Affairs, 2012, bitly.com/oil-revolution).
2. 'Point of No Return: The Massive Climate Threats We Must Avoid' (Greenpeace, 2013, bitly.com/carbon-bomb).
3. BP states that, 'Our "Policy Case" … assumes a step-change in the political commitment to action on carbon emissions. Even in this case, the path to reach 450 ppm remains elusive. However, a declining emissions path by 2030 is achievable, given the political will to shoulder the cost.' We requested an interview with BP to discuss these issues but they didn't respond.
4. Our $100 rough figure is based on an average at the time of writing between Brent crude ($112) and West Texas Intermediate ($89), as listed on bloomberg.com/energy. Oil reserves figure from BP's 'Statistical Review of World Energy 2012' (bp.com/statisticalreview).

5. Arriving at an appropriate discount rate is not by any means an exact science. Four per cent is the figure suggested by the UK's Office of National Statistics for estimating the value of the UK's oil reserves. They write: 'This rate can be approximated by the interest rate on low risk bonds. The choice of discount rate is a very debatable issue as it involves balancing considerations of intergenerational equity (the discount rate should be equal to zero in this case), social time preference, the social opportunity cost of capital and (in the case of oil and gas) the degree to which fossil fuels will be substitutable in the future or will prove to have other economic uses. The precise value [of the social discount rate] is fairly arbitrary but a figure of 4 per cent has been considered acceptable by the Eurostat Task Force.' Source: 'The Valuation of Oil and Gas Reserves' (ONS, 1998, bitly.com/discount-rate).

6. 'Unburnable Carbon' (Carbon Tracker, 2013, being drafted at the time of writing). Interestingly and encouragingly, this total market capitalisation is more than a third lower than it was in same report in February 2011. It isn't clear how much of this loss of value (if any) is due to increased investor concern about the potential impact of climate legislation on fuel company profits.

7. Personal correspondence with James Leaton of Carbon Tracker.

8. The IEA estimates that approximately two-thirds of proven coal reserves are government owned, and around 90 per cent of proven oil and gas reserves. Source: *World Energy Outlook 2012*, Figure 8.12: Potential CO_2 emissions from remaining fossil-fuel reserves by fuel type.

9. These rough figures are based on the fossil fuel activity of listed coal companies being worth $0.7 trillion (and accounting for 35 per cent of all coal reserves) and the listed oil and gas sector being worth $3.3 trillion (and accounting for 10 per cent of all oil and gas reserves).

10. A Deutsche Bank assessment estimated the average cash costs of the major producers are $7.70 a barrel, with the UAE, Kuwait and Saudi Arabia coming in below $2. Source: 'European Integrated

Oils' (Deutsche Bank, 2009, available at the time of writing at bitly.com/oil-costs). By contrast one recent analysis put the cost of production for BP at close to $40 and suggested that the marginal cost (the cost of increasing production by one barrel) is closer to $100. Source: Kate Mackenzie, 'Marginal Oil Production Costs Are Heading towards $100/Barrel' (FT.com, 2012, bitly.com/extraction-costs).

11. 'The market value of Nasdaq companies peaked at $6.7 trillion in March 2000 and bottomed out at $1.6 trillion in October 2002.' Gaither and Chmielewski, 'Fears of Dot-Com Crash, Verson 2.0', *Los Angeles Times*, 16 July 2006. Available at: bitly.com/dot-com-crash.

12. BP's annual 'Statistical Review of World Energy' is a standard source for energy reserve data. The latest version is made available at: bp.com/statisticalreview. This book was written using figures from the 2012 edition, though the map is based on the 2011 edition, in which Venezuelan reserves appear smaller.

13. Map from CarbonMap.org, created by Duncan Clark and Robin Houston, based on BP's 'Statistical Review of World Energy 2011'. Note that the BP figures only list countries with large reserves, with the remainder listed under 'Other Africa', 'Other Asia/Pacific'. These unallocated emissions make up a few per cent of the estimated total. For this map, the unallocated reserves for each region have been distributed between unlisted countries relative to their surface area.

14. The Cartagena Dialogue is a loose-knit body with no formal membership list. We've taken the membership to be as follows: Antigua & Barbuda, Australia, Bangladesh, Belgium, Burundi, Chile, Colombia, Cook Islands, Costa Rica, Democratic Republic of the Congo, Denmark, Dominican Republic, Ethiopia, European Union, France, Gambia, Germany, Ghana, Grenada, Guatemala, Indonesia, Kenya, Lebanon, Lesotho, Malawi, Maldives, Marshall Islands, Netherlands, New Zealand, Norway, Panama, Peru, Poland, Rwanda, Samoa, Spain, Sweden, Switzerland, Tajikistan,

Tanzania, Thailand, Timor-Leste, United Arab Emirates, United Kingdom and Uruguay. This list is based on various press releases and interviews carried out by environmental studies student Cecilia Pineda (bitly.com/pineda-2012). We've excluded Mexico and South Africa who are marked as observers. Based on this list, our calculations suggest that EU, Cartagena Dialogue, AOSIS and Africa collectively control 19.8 per cent of the carbon in proven fossil fuel reserves.

15. Ratio of reserves to production from the BP 'Statistical Review of World Energy 2012' (bp.com/statisticalreview).

16. Based on graph from IEA, *World Energy Outlook 2012*, Figure 8.12, Potential CO_2 emissions from remaining fossil-fuel reserves by fuel type.

17. For example, the UK stock market in 2010 was owned 41 per cent by foreign investors, 9 per cent by insurance companies, 5 per cent by pension funds (a historic low), 16 per cent by other financial institutions and 12 per cent by UK individuals. Source: 'Statistical Bulletin: Ownership of UK Quoted Shares' (Office for National Statistics, 2010, bitly.com/ons-shares)

18. 'Unburnable Carbon: Are the World's Financial Markets Carrying a Carbon Bubble?' (Carbon Tracker, 2012, bitly.com/carbonbubble).

19. HSBC Global Research, 'Oil & Carbon Revisited: Value at Risk from "Unburnable" Reserves' (HSBC, 2013).

20. The HSBC analysis is based on IEA 450ppm scenario which includes major roll-out of CCS, as shown in the chart in chapter ten of this book. The scenario assumes climate change is tackled by national commitments to cut carbon, which force down demand with the effect of reducing the global trading price, even as end-user prices rise due to carbon. If instead there was a global cap-and-trade system on extraction then the trading price might be high but the profits would be reduced instead by the cost of acquiring permits.

21. Figures from opensecrets.org/industries, as of February 2013. The numbers for the alternative energy sector were $2.5 million in donations and $28 million in lobbying. Of the fossil fuel money, more than three-quarters has gone to the Republicans, though a sizable amount was received by politicians of other parties, including President Obama himself, who is reported to have taken $884,000 dollars from oil and gas companies in the run-up to the 2008 election (source: bitly.com/os-oil-gas).

22. At the Sleipner project in the Norwegian North Sea, carbon dixoide is injected back into a gas well to avoid the national tax on carbon emissions. There are no signs of leakage, according to studies such as: R. Arts *et al.*, 'Recent Time-lapse Seismic Data Show No Indication of Leakage at the Sleipner CO_2-injection Site' (GHGT-7, 2004). For more sites, see Hosa *et al.*, 'Benchmarking worldwide CO_2 saline aquifer injections' (Scottish Centre for Carbon Capture and Storage, 2010, bitly.com/hosa-2010).

23. For an excellent summary of CCS technologies, see the briefing paper from the Grantham Research Institute at Imperial College. Separate papers for capture and storage can be found via: bitly.com/icl-publications.

24. The first flagship CCS scheme in the US, launched by President George W. Bush in 2003, was FutureGen – a plan for a zero-emissions coal plant in Mattoon, Illinois. After various setbacks, government money was finally pulled in 2008. After President Obama came to office, the project was revived under the name FutureGen 2.0 (see: bitly.com/futuregen-2). In the UK, CCS support began with a government-run competition. The scheme was widely criticised and reached crisis-point in late 2011 when the only remaining contender dropped out. See: bitly.com/guardian-longannet.

25. The Global CCS Institute maintains an up-to-date list of CCS projects on its website at: bitly.com/ccs-projects.

26. M. Blunt, 'Grantham Institute for Climate Change Briefing Paper No. 4: Carbon Dioxide Storage' (2010, bitly.com/ccs-volume).

27. N. McGlashan *et al.*, 'Grantham Institute for Climate Change Briefing Paper No. 8: Negative Emissions Technologies' (2012, bitly.com/negative-carbon).

28. The $95/tonne figure is from the Grantham paper cited in the previous note, based on Lackner's own figures. However, Lackner has also predicted lower long-term costs. For example, at a talk at Imperial College London in September 2009 he said that he believes $30–40 may eventually be realistic.

29. A very readable history of Lackner's efforts to create carbon scrubbers is provided in R. Kunzig and W. Broecker, *Fixing Climate: The Story of Climate Science and How to Fix Global Warming* (Profile Books, 2009).

30. An excellent overview of each approach is given in Chris Goodall's book *Ten Technologies to Fix Energy and Climate* (Profile Books, 2009).

31. A good introduction to Savory's approach (complete with lots of photos of case studies) is his 2013 TED talk, 'How to Green the World's Deserts and Reverse Climate Change', available at bitly.com/savory-ted.

32. 'Geoengineering the Climate: Science, Governance and Uncertainty' (Royal Society, 2009, bitly.com/rs-geoengineering).

33. This claim is made in Savory's TED talk, bitly.com/savory-ted, though at the time of writing that

34. A. Savory, 'A Global Strategy for Addressing Global Climate Change' (undated, bitly.com/savory-climate).

35. As of mid-2012, the website of the Organisation Internationale des Constructeurs d'Automobiles claims the global automotive industry turns over €1,889,840 billion, which equates to around $2.3 trillion. Source: oica.net/economic-facts. World vehicle registrations in 2010 came to 1.015 billion, excluding off-road vehicles, according to Ward's Auto (bitly.com/wards-auto).

36. IEA, *World Energy Outlook 2012*, chapter 8: Carbon in Energy Reserves and Energy Infrastructure.

37. S. Davis *et al.*, 'Future CO_2 Emissions and Climate Change from Existing Energy Infrastructure' (*Science*, 2010, bitly.com/davis-2010).

38. A. Yang and Y. Cui, 'Global Coal Risk Assessment: Data Analysis and Market Research' (WRI, 2012, bitly.com/wri-coal-risk). Our back-of-the-envelope calculation runs as follows. The WRI report found 1,401,278 megawatts of planned coal capacity. Assuming typical usage of 7,000 hours a year, that's 10 billion megawatt hours of power a year, or 350 billion megawatt hours over a lifetime of 35 years. At present, a megawatt hour of coal power in the UK produces around 910kg of carbon dioxide per megawatt hour. If we round that figure down to, say, 800kg to reflect efficiency gains in power stations, that would still give us lifetime emissions of 280 billion tonnes of CO_2. That's more than a third of the remaining carbon budget for a 75 per cent chance of limiting global warming to 2°C.

39. Figures from the Organisation Internationale des Constructeurs d'Automobiles, which publishes statistics by nation and year. Available at: bitly.com/oica-production.

40. Davis *et al.* (see note 2, Chapter 9) put the figures at '17, 16, and 28 years for passenger cars, light trucks, and heavy vehicles (trucks and buses) respectively', based on figures from the US, while 'coal, natural gas, and oil [power plants run for] 38.6, 35.8, and 33.8 years, respectively.'

41. Based on comments by Li Jiaxiang, director of the Civil Aviation Administration of China, quoted in L. Dongmei, '70 New Airports Planned, Aviation Official Says' (Caixin Online, 2012, bitly.com/china-airports).

42. Such a process is possible: with sufficient clean electricity, hydrogen can be electrolysed from water; CO_2 can be captured from air; and the two can be combined to produce synthetic liquid fuels that are effectively carbon neutral. This is unlikely ever

to be cost-competitive with electric cars but in theory could be useful for planes and other vehicles and devices that require liquid fuels in a carbon constrained world. One British firm exploring this technology is described in A. Hough, 'British Engineers Produce Amazing "Petrol from Air" Technology' (*Telegraph*, 2012, bitly.com/air-fuel).

Chapter 9: The growth debate

1. This breakdown was created by Mike for campaign group 10:10. All emissions are allocated to the end consumer, so 'paper and printing' for instance includes not just the fuels burned to harvest trees and process pulp, but also the laptops and business flights of staff working in that sector and even the paper industry's share of the emissions caused by financial and legal companies. This approach is called 'input-output' carbon footprinting and forms the basis of much of Mike's professional work. Another breakdown of the same numbers is shown in the chart below.

2. The company – Planetary Resources – launched to great media fanfare in spring 2012. Its investors include Google's Eric Schmidt and Larry Page, and *Avatar* director James Cameron, among others. See: planetaryresources.com.

3. Obama's first extended statement on climate change in a long period was made on 14 November 2012, just after both the election and Hurricane Sandy. He said: 'If the message is somehow we're going to ignore jobs and growth simply to address climate change I don't think anybody's going to go for that. I won't go for that. If on the other hand we can shape an agenda which says we can create jobs, advance growth and make a serious dint in climate change and be an international leader, then I think that's something the American people would support.' Currently available at: bitly.com/obama-growth.

4. *Stern Review on the Economics of Climate Change* (HM Treasury, 2006, bitly.com/stern-review).

5. K. Anderson and A. Bows, 'Beyond Dangerous Climate Change: Emission Scenarios for a New World' *(Philosophical Transactions of the Royal Society*, 2011, bitly.com/beyond-dangerous).

6. For Tim Jackson's perspective, see P. Victor and T. Jackson, 'It's Not Easy Being Green' (*Dimensions*, 2012, bitly.com/not-easy-green). The piece concludes that green growth would require 'completely unprecedented, almost certainly unrealistic, levels of improvement in technological efficiency'. See also Jackson's book, *Prosperity without Growth* (Earthscan, 2009) or his report of the same title for the UK government's Sustainable Development Commission (bitly.com/prosperity-growth).

7. IEA, *World Energy Outlook 2011*.

8. The *Stern Review* made its own decision over this, which was contentious but arguably no less defensible than the alternatives. Stern picked a 'social discount rate' of 1 per cent per year, based loosely on the idea that the lives of our grandchildren are as important as our own but there is a possibility that whatever we do, they might not get to exist anyway – in which case efforts to improve their future are wasted.

9. R. Wilkinson and L. Pickett's *The Spirit Level* (Penguin, 2010), looks at many of these shortcomings in an accessible way. For more on the economic limitations of GDP, see J. Stiglitz *et al.*, 'The Report of the Commission on the Measurement of Economic Performance and Social Progress' (2009, bitly.com/social-progress).

10. It's not entirely clear who first said that what gets measured gets done. The phrase is widely associated with people such as such as Tom Peters and Peter Drucker but the sentiment can be traced back at least to Lord Kelvin, who wrote in 1883: 'When you can measure what you are speaking about, and express it in numbers, you know something about it; but when you cannot measure it, when you cannot express it in numbers, your knowledge is of a meagre and unsatisfactory kind; it may be the beginning of knowledge, but you have scarcely in your thoughts advanced to

the stage of science.' Source: Sir William Thompson, *Popular Lectures and Addresses Volume I* (1889).

11. The Gross National Happiness website (bitly.com/national-happiness) sums up the metric as follows: 'Gross National Happiness is a term coined by His Majesty the Fourth King of Bhutan, Jigme Singye Wangchuck in the 1970s. The concept implies that sustainable development should take a holistic approach towards notions of progress and give equal importance to non-economic aspects of wellbeing. The concept of GNH has often been explained by its four pillars: good governance, sustainable socio-economic development, cultural preservation, and environmental conservation. Lately the four pillars have been further classified into nine domains in order to create widespread understanding of GNH and to reflect the holistic range of GNH values. The nine domains are: psychological wellbeing, health, education, time use, cultural diversity and resilience, good governance, community vitality, ecological diversity and resilience, and living standards. The domains represent each of the components of wellbeing of the Bhutanese people, and the term "wellbeing" here refers to fulfilling conditions of a "good life" as per the values and principles laid down by the concept of Gross National Happiness.'

12. J. Stiglitz *et al.*, 'The Report of the Commission on the Measurement of Economic Performance and Social Progress' (2009, bitly.com/social-progress).

13. 'How the country is doing' is the phrase used by the ONS in one of their launch documents for the project, 'Measuring what Matters: National Statistician's Reflections on the National Debate on Measuring National Well-being' (ONS, 2011, bitly.com/ons-well-being).

14. In 'The Sustainable Borders of the State' (*Oxford Review of Economic Policy*, 2011, bitly.com/helm-state), Dieter Helm argues that natural resources should be treated as assets that can create wealth.

15. For Cameron's 'every department to be a growth department' comment, see P. Wintour, 'Cross-party Heathrow Runway Talks to Begin' (*The Guardian*, 2012, bitly.com/heathrow-runway). For background on removal of transport secretary Justine Greening, see: 'Cameron Reshuffles His Government: Politics Live Blog' (*The Guardian*, 2012, bitly.com/cameron-reshuffle). For the G7 call on OPEC see 'G7 Urges Oil Supply Boost' (Platts, 2012, bitly.com/g7-opec). The statement was made by the group's finance ministers and said: 'We encourage oil-producing countries to increase their output to meet demand, while drawing prudently on excess capacity.'

Chapter 10: The great global slumber

1. For a fascinating and highly readable summary of Daniel Kahneman's work on various types of cognitive biases, read his bestseller *Thinking Fast and Slow* (Penguin, 2012).
2. T. Sharot, *The Optimism Bias: Why We're Wired to Look on the Bright Side* (Pantheon Books, 2011). An extract is available at bitly.com/opt-bias.
3. R. Dawkins, 'Sustainability Doesn't Come Naturally: A Darwinian Perspective on Values', Inaugural Lecture of the Values Platform for Sustainability (The Environment Foundation, 2001, bitly.com/dawkins-talk).
4. C. Weeramantry *et al.*, 'Guardians for the Future: Safeguarding the World from Environmental Crisis' (*The Guardian*, 2012 bitly.com/wfc-idea).
5. Goebbert *et al.*, 'Weather, Climate, and Worldviews: The Sources and Consequences of Public Perceptions of Changes in Local Weather Patterns'. (*Weather, Climate and Society,* 2012, bitly.com/weather-perception).
6. Quoted in: C. Rapley, 'Climate Science: Time to Raft up'. (*Nature*, 2012, bitly.com/ramp-up).

7. Leiserowitz runs the Yale Project on Climate Change Communication. This quote is from B. Gardiner, 'We're All Climate-Change Idiots' (*New York Times*, 2012, bitly.com/climate-idiots).

8. N. Oreskes and E. Conway, *Merchants of Doubt: How a Handful of Scientists Obscured the Truth on Issues from Tobacco Smoke to Global Warming* (Bloomsbury Press, 2010). For a summary of much of the book's content, and a good background on the long history of the science of anthropogenic global warming, watch Oreskes' Vetlesen Lecture at the University of Rhode Island (2010, bitly.com/oreskes-talk).

9. Papers published in 2005 showed that some years earlier US under-secretary of state, Paula Dobriansky, had written to GCC saying that President George W. Bush had 'rejected Kyoto in part based on input from you'. Source: J. Vidal, 'Revealed: How Oil Giant Influenced Bush' (*The Guardian*, 2005, bitly.com/gcc-bush).

10. The GCC's website, after it shut down, said: 'The Global Climate Coalition has been deactivated. The industry voice on climate change has served its purpose by contributing to a new national approach to global warming. The Bush administration will soon announce a climate policy that is expected to rely on the development of new technologies to reduce greenhouse emissions, a concept strongly supported by the GCC.' The website can still be accessed via Internet Archive at: bitly.com/gcc-archive.

11. You can easily find the commercials online, for example at: bitly.com/cei-co2. The CEI's funders that year included ExxonMobil and the American Petroleum Institute according to J. Achenbach, 'The Tempest' (*Washington Post*, 2006, bitly.com/cei-funders).

12. The quote is taken from a Heartland Institute press release, available at: bitly.com/heartland-press. This particular campaign was so radical that many sponsors abandoned the organisation, as reported in S. Goldenberg, 'Big Donors Ditch Rightwing Heartland Institute over Unabomber Billboard' (*The Guardian*, 2012, bitly.com/heartland-exodus).

13. The memo was leaked in 2003. See: O. Burkeman, 'Memo Exposes Bush's New Green Strategy' (*The Guardian*, 2003. bitly.com/bush-strategy).

14. Inhofe's book about climate change is called *The Greatest Hoax* (WND Books, 2012). The comment about Nazis came in an interview with Tulsa World in which he compares the push to reduce emissions with the Third Reich (see bitly.com/inhofe-interview). The figure for his income from oil and gas companies is taken from the Open Secrets database and covers his career since 1989 (see bitly.com/inhofe-funds).

15. Sean Hannity quote from the video available at bitly.com/hannity-video. Glenn Beck quote from 'Global Warming's Real Inconvenient Truth' (Fox News, 2009, bitly.com/beck-foxnews).

16. C. Meredith, '100 Reasons Why Climate Change Is Natural' (*Daily Express*, 2012, bitly.com/climate-natural).

17. Union of Concerned Scientists, 'Got Science? Not at News Corporation' (2012, bitly.com/climate-science).

18. Research carried out by CCR Group, looking at coverage in July 2012 in *The Sun*, *The Times*, *Daily Telegraph*, *Daily Mail* and *Daily Mirror*. Available at 'CC Group Finds National Media "Anti-renewables" & Neglecting Industry voice' (bitly.com/media-anti). This survey concurs with another carried out in 2009 by the Public Interest Research Centre, summarised in: D. Clark, 'How UK Newspaper Coverage Is Skewed against Renewables' (*The Guardian*, 2012, bitly.com/media-renewables).

19. G. Monbiot, 'Top Gear's Electric Car Shows Pour Petrol over the BBC's Standards' (*The Guardian*, 2011, bitly.com/monbiot-topgear).

20. E. Lipton and C. Krauss, 'Fossil Fuel Industry Ads Dominate TV Campaign' (*The New York Times*, 2012, bitly.com/climate-ads).

21. Various enquiries later cleared them of any scientific misconduct. The most in-depth was 'The Report of the Independent Climate Change Email Review' (2010, cce-review.org).

22. 'King's College London: Global Advisor Questions' (Ipsos MORI, 2010, bitly.com/ipsos-questions).

23. 'Obama More Popular Abroad Than at Home, Global Image of U.S. Continues to Benefit' (Pew Research Center, 2010, bitly.com/pew-climate).

24. 'Individual Perceptions of Climate Risks Survey' (AXA/Ipsos, 2012, bitly.com/axa-perceptions).

25. Although recent experience in the US suggests that extreme weather events do indeed raise the profile of climate change, it's not by any means a given that the effects always work as expected. As climate communication specialist George Marshall has noted, such events may sometimes have unexpected side effects – such as encouraging 'powerful and compelling survival narratives that can overwhelm weaker and more complex climate change narratives'. Source: G. Marshall, 'Will Hurricane Sandy Increase Concern about Climate Change?' (Talking Climate, 2012, bitly.com/weather-response).

26. J. Eilperin and P. Craighill, 'Global Warming no Longer Americans' top environmental concern, poll finds' (*The Washington Post*, 2012, bitly.com/cliamte-survey).

27. F. Newport, 'Americans Endorse Various Energy, Environment Proposals' (Gallup Politics, 2012, bitly.com/newport-gallup). For a list of other polls showing US public support for action see: J. Romm, 'Gallup: 65 per cent of Americans Have More Guts Than Obama, Support "Imposing Mandatory Controls on CO_2 Emissions"' (Think Progress, 2012, bitly.com/think-polls).

28. Leiserowitz *et al.*, 'Extreme Weather and Climate Change in the American Mind' (Yale Project on Climate Change Communication, 2012, bitly.com/leiserowitz-2012).

29. The cover article is P. Barrett, 'It's Global Warming, Stupid' (*Bloomberg Business Week*, 2012, bitly.com/warming-stupid). The cover itself is easily located on Google Images.

30. The experiment was conducted for TV in 2001 by Stanford professor Philip Zimbardo. The subjects 'remained in the room for an average of 13 minutes doing nothing, despite billowing smoke and ringing alarms' according to a university press release available at: bitly.com/zimbardo-smoke.

31. D. Kahan, 'Why We Are Poles Apart on Climate Change' (*Nature*, 2012, bitly.com/kahan-nature).

32. This tendency, which they dubbed the 'availability heuristic', is written up in D. Kahneman, *Thinking Fast and Slow* (Penguin, 2012).

33. The 'iron cage of consumerism' is a reference by Tim Jackson (*Prosperity Without Growth*, Earthscan, 2009) to Max Weber's 'Iron Cage of Bureaucracy' in *The Protestant Ethic and the Spirit of Capitalism* (translated into English 1930).

34. 'Think Of Me As Evil' (PIRC, 2011, bitly.com/pirc-think). This interesting report includes a wealth of links to research exploring the various issues around advertising, consumption and values. The lack of academic evidence on advertising driving overall consumption chimes with our own attempts to map a correlation between the amount spent on advertising in a country and the carbon footprint of its citizens. Plotting advertising spend as a percentage of GDP (from: *The Spirit Level* based on World Advertising Center, 'World Advertising Trends', 2002) against total carbon footprints in each country, including imported goods (from: G. Peters *et al.*, 'Rapid Growth in CO_2 Emissions after the 2008–2009 Global Financial Crisis', *Nature Climate Change*, 2012, bitly.com/peters-2012), there is no correlation across OECD nations and if anything a negative correlation across Europe – though not one that looks statistically significant.

35. Cited in the PIRC report, from 'We Can't Run Away from the Ethical Debates in Marketing' (Market Leader, 2010)

36. Income inequality data based on the *Spirit Level* dataset, which draws on averages from UN Development Program, Human Development Report, 2003, 2004, 2005, 2006. Carbon footprint data from G. Peters *et al.*, 'Rapid Growth in CO_2 Emissions after the 2008–2009 Global Financial Crisis' (*Nature Climate Change*, 2012, bitly.com/peters-2012). We found no meaningful relationship between the two. It may be possible to argue that there is a very slight correlation across OECD nations, but it isn't statistically significant. Moreover, when just (more comparable) European nations are looked at, there is a stronger trend in the opposite direction.

37. M. Ravallion *et al.*, 'Carbon Emissions and Income Inequality', *Oxford Economic Papers*, 2000, bitly.com/ravallion-2000). This study suggested a positive correlation between higher economic equality and higher carbon footprints. It doesn't draw on data that takes account of imported goods, so it's findings are of limited value, but the paper does add more weight to the idea that reducing inequality wouldn't in itself reduce carbon footprints.

38. This claim draws data from Wilkinson *et al.*, 'Equality, Sustainability, and Quality of Life' (*British Medical Journal*, 2010, bitly.com/wilkinson-2010).

Chapter 11: The problem of sharing

1. G. Hardin, 'The Tragedy of the Commons' (*Science*, 1968, bitly.com/tragedy-commons).

2. C. Goodall, 'The Human Brain Is Made for Environmental Complacency' (Carbon Commentary, 2009, bitly.com/cc-brain).

3. One early advocate of regulating carbon emissions 'upstream' in this way was Oliver Tickell in his Kyoto 2 framework and book. For more information see: kyoto2.org.

4. SAFE stands for 'Sequestered Adequate Fraction of Extracted'. The idea was proposed in M. Allen *et al.*, 'The Case for Mandatory Sequestration' (*Nature Geoscience,* 2009, bitly.com/safecarbon).

5. S. Greenfield, *ID: The Quest for Identity in the Twenty-First Century* (Sceptre, 2008).

Chapter 12: The supporting cast

1. The data behind the chart is as follows:

World Greenhouse Gas Emissions
Gt CO_2e per year. Comparisons based on a century timeframe

Carbon dixoide from fossil fuels	**31,874**	**65.9%**
Coal	14,343	29.7%
Oil	11,156	23.1%
Gas	6375	13.2%
Carbon dioxide from deforestation	**3190**	**6.6%**
Food	1914	4.0%
Timber	1276	2.6%
Carbon dixoide from cement production	**1639**	**3.4%**
Methane	**6875**	**14.2%**
Fossil fuel burning	2530	5.2%
Enteric emissions	1891	3.9%
Manure	213	0.4%
Rice cultivation	701	1.5%
Burning	413	0.9%
Waste disposal	1134	2.3%
Other	7	0.0%
Nitrous oxide	**3159**	**6.5%**
Fossil fuels	272	0.6%
Chemical production	373	0.8%
Manure	1482	3.1%
Indirect from agriculture	265	0.5%
Burning	423	0.9%
Waste	111	0.2%
Indirect from non-agriculture	212	0.4%
F-gases	**904**	**1.9%**
Aviation additional effects	**698**	**1.4%**
Total	**48,339**	

Sources: Fossil fuel, cement and other industries and land use change CO_2 emissions data from: T. Boden *et al.*, 'Global, Regional, and National Fossil-Fuel CO_2 Emissions' and R. Houghton 'Carbon Flux to the Atmosphere from Land-Use Changes: 1850–2005' (both Carbon Dioxide Information Analysis Center, 2010), and G. Peters *et al.*, 'Rapid Growth in CO_2 Emissions after the 2008–2009 Global Financial Crisis' (*Nature Climate Change*, 2012).

The split in fossil fuel emissions comes from International Energy Agency and describes 2011 (IEA, 2012, bitly.com/iea-co2-increase). The totals and splits of other gases come from the European Commission's Emission Database on Global Atmospheric Research (EDGAR, 2008, bitly.com/edgar-research).

The drivers of deforestation are difficult to attribute in a definitive quantified way. We have split them 40 per cent timber and logging, 60 per cent agriculture based on analysis in Mongabay.com (bitly.com/mongabay-forest). This fits with a WWF estimate that 58 per cent of deforestation is due to commercial agriculture. Audsley *et al.*, 'How Low Can We Go? An Assessment of Greenhouse Gas Emissions from the UK Food System and the Scope to Reduce by 2050' (FCRN/WWF, 2009, bitly.com/wwf-how-low).

2. This was the conclusion in: T. Bond *et al.*,'Bounding the Role of Black Carbon in the Climate System: A Scientific Assessment' (*Journal of Geophysical Research: Atmospheres*, 2013, bitly.com/soot-paper). It found that: 'black carbon [is] the second most important human emission in terms of its climate-forcing in the present-day atmosphere'.

3. Methane's total impact is longer-lived as it breaks down to form CO_2. But this long-term effect is small compared to its intense short-term effect.

4. For a layman-friendly overview on the science and uncertainty around global dimming see: The Met Office and D. Clark,

'What Is Global Dimming?' (*The Guardian*, 2012, bitly.com/global-dimming).

5. One recent paper projected the timing of temperature increases based on IPCC emissions scenarios. In the scenarios closest to what's happened so far (A1B and A2) over 90 per cent of projections give 2°C within fifty years and a high danger of 3°C. Source: Joshi *et al*, 'Projections of When Temperature Change Will Exceed 2°C above Pre-industrial Levels' (*Nature Climate Change*, 2011, bitly.com/when-2c).

6. 'Integrated Assessment of Black Carbon and Tropospheric Ozone: Summary for Decision Makers' (UNEP and WMO, 2011, bitly.com/black-carbon).

7. J. Blackstock and M. Allen, 'Oxford Martin Policy Brief: The Science and Policy of Short-Lived Climate Pollutants' (Oxford Martin School, 2012, bitly.com/slcp-effect).

8. This process is being developed by Novacem (bitly.com/novacem).

9. A. Macintosh and L. Wallace, 'International Aviation Emissions to 2025: Can Emissions Be Stabilised without Restricting Demand?' (*Energy Policy*, 2009, bitly.com/flying-demand). Industry estimates predict that revenue passenger kilometres (RPK) will increase at an average rate of 5 per cent per year for the next twenty years, and cargo traffic by 6 per cent over the same period. These forecasts would equate to an increase in passenger RPK of 180 per cent in the period 2006–2026 and an increase in cargo RPK of 220 per cent.

10. A. Hough, 'British Engineers Produce Amazing "Petrol from Air" Technology' (*Daily Telegraph*, 2012, bitly.com/air-fuel).

11. D. Clark, 'Could Computerising Air-traffic Control Save Carbon, Time and Money?' (*The Guardian*, 2011, bitly.com/aviation-traffic).

12. For a summary of the physical limits, see David MacKay, *Sustainable Energy: Without the Hot Air* (UIT, 2008).

13. Based on Civil Aviation Authority data survey of 6500 passengers modelled by Small World Consulting with technical assistance from David Parkinson of Sensus UK. All respondents were residents of Greater Manchester or West Sussex and we assume those two locations tolerably represent the UK. A map and summary are available at: D. Clark, 'The Carbon Footprint of British Holiday Flights – interactive' (*The Guardian*, 2012, bitly.com/air-travel-carbon).

Chapter 13: Food, forests and fuels

1. There were an estimated 963 million undernourished people in the world in 2009, out of a population that was 6.76 billion at the time. Source: Nellemann *et al.*, 'The Environmental Food Crisis: The Environment's Role in Averting Future Food Crises' (UNEP/GRID-Arendal, 2009, bitly.com/unep-food-crisis).
2. The quote is from Professor Carsten Rahbek, Director for the Center for Macroecology, Evolution and Climate at the University of Copenhagen. Available at: bitly.com/biodiversity-crisis. Rahbek's view, widely shared in the scientific community, is that thanks to human activity, the Earth is currently in the throes of a sixth period of mass extinction. See also: Millennium Ecosystem Assessment, 'Ecosystems and Human Well Being: Biodiversity Synthesis' (UNEP, 2005, bitly.com/unep-biodiversity) and Barnosky *et al.*, 'Has the Earth's Sixth Mass Extinction Already Arrived?' (*Nature*, 2011, bitly.com/sixth-me). The latter paper treats the biodiversity in a more mathematical way and compares it to the 'big five' mass extinctions while accepting that this is difficult due to incompleteness of the fossil record and other data. It concludes that while the current crisis is not yet of the scale and rapidity of the previous five, the extinction rate is still worryingly high and deserves rapid action.
3. Deforestation is the main cause of emissions related to land use change. For our purposes the two are more or less interchangeable.

4. One recent paper said that by 2008 deforestation was 12 per cent of total CO_2, or 15 per cent if peatlands are included. These proportions are likely to have fallen since given the continued rise of fossil fuel use. Source: G. van der Werf *et al.*, 'CO_2 Emissions from Forest Loss' (*Nature Geoscience*, 2009, bitly.com/van-der-werf).

5. The figure of 13 million hectares is from the UN Food and Agriculture Organization (bitly.com/fao-forest-assess).

6. The United Nations Environment Programme reports that: 'It takes, on average, 3 kg of grain to produce 1 kg of meat, given that part of the production is based on other sources of feed, rangeland and organic waste … Hence, an increased demand for meat results in an accelerated demand for water, crop and rangeland area.' Source: Nellemann *et al.*, 'The Environmental Food Crisis – The Environment's Role in Averting Future Food Crises' (UNEP/GRID-Arendal, 2009, bitly.com/unep-food-crisis).

7. Seventy per cent of previously forested land in the Amazon is now livestock pasture with most of the remainder used for animal feed. Steinfeld *et al.*, 'Livestock's Long Shadow' (UN FAO, 2006, bitly.com/long-shadow).

8. P. McMahon, *Feeding Frenzy* (Profile Books, 2013).

9. *OECD – FAO Agricultural Outlook 2012–2021* (bitly.com/oecd-outlook).

10. Official recent statistics from Brazil are available at: bitly.com/brazil-stats. For a summary of changes since 2004, see: A. Vaughan, 'Amazon Deforestation Falls Again' (*The Guardian*, 2012, bitly.com/am-def). For an engaging write-up of the work of Brazil's rapid response teams, see J. Watts, 'Brazil's Amazon Rangers Battle Farmers' Burning Business Logic' (*The Guardian*, 2012, bitly.com/amazon-rangers).

11. M. Berners-Lee, *How Bad are Bananas? The Carbon Footprint of Everything* (Profile Books, 2010). Throughout this book there

is quite a bit on the carbon footprint of food and advice for shoppers.

12. All estimates apart from those for rice come from two studies by Williams *et al.* at Cranfield University, commissioned by the UK's Department for Agriculture and Rural Affairs. Figures for beef, lamb, chicken, and potatoes come from 'Comparative Life-cycle Assessment of Food Commodities Procured for UK Consumption Through a Diversity of Supply Chains' (2008, bitly.com/defra-2008). Figures for pork, eggs and wheat flour come from 'Determining the Environmental Burdens and Resource Use in the Production of Agricultural and Horticultural Commodities' (2006, bitly.com/defra-2006). Figures for rice are taken from: Kasmaprapruet *et al.*, 'Life-cycle Assessment of Milled Rice Production: Case Study in Thailand' (*European Journal of Scientific Research*, 2009). These and many more emissions factors for foods can be found collated in Small World Consulting's work for Booths supermarkets: 'Booths Greenhouse Gas Footprint Report 2012' (bitly.com/booths-report).

13. *Climate Change 2007: Synthesis Report*, 3.2.2: Impacts on regions (IPCC, 2007, bitly.com/ipcc-food).

14. Lobell *et al.*, 'Climate Trends and Global Crop Production since 1980'. (*Science*, 2011, bitly.com/climate-crops),

15. Hawkins *et al.*, 'Increasing influence of heat stress on French Maize Yields from the 1960s to the 2030s' (*Global Change Biology*, 2012). The study concludes that 'to offset the projected increased daily maximum temperatures over France, improved technology will need to increase base level yields by 12 per cent to be confident about maintaining current levels of yield for the period 2016–2035; the current rate of yield technology increase is not sufficient to meet this target'.

16. W. Cheung *et al.*, 'Shrinking of Fishes Exacerbates Impacts of Global Ocean Changes on Marine Ecosystems' (*Nature Climate Change*, 2012, bitly.com/shrinking-fish).

17. Hansen *et al.*, 'Public Perception of Climate Change and the New Climate Dice' (PNAS, 2012, bitly.com/hansen-2012). See chapter two for more details.

18. OECD–FAO *Agricultural Outlook 2012* (bitly.com/oecd-outlook).

19. 9.2 KWhr per litre of biodiesel is 7823kCal.

20. The table opposite is based on figures from the Biomass Energy Centre, UK (biomassenergycentre.org.uk) with the right hand column added by us, based on 10.4 KWh per litre of mineral diesel.

21. Numbers taken from J. Lundqvist *et al.*, 'Saving Water: From Field to Fork – Curbing Losses and Wastage in the Food Chain' (SIWI Policy Brief, 2008, bitly.com/field-to-fork), which in turn quotes them from V. Smil, *Feeding the World: A Challenge for the Twenty-First Century* (MIT Press, 2001). Cross-checked them against various other sources.

22. This is a back-of-the-envelope calculation based on World Bank data showing that wheat yields around the world mainly range from 2000 to 7000 kg per hectare (source: bitly.com/cereal-yield). We've assumed a calorific value of 3400 kCal per kilogram and just over 2000 calories required per person per day.

23. Anseeuw *et al.* (2011), 'Land Rights and the Rush for Land'. International Land Coalition. Available at: bitly.com/rush-for-land.

24. OECD–FAO *Agricultural Outlook 2012* (bitly.com/oecd-outlook).

Chapter 14: Waking up

1. Quote from speech reported by John Parnell in 'IEA: Business Complacent about Climate Change' (RTCC.org, 2013, bitly.com/4C-ok).

2. S. Goldenberg, 'Revealed: the day Obama chose a strategy of silence on climate change' (*The Guardian*, 2012. bitly.com/obama-silent).

Fuel	Net calorific value (MJ/kg)	Annual yield per hectare (tonnes)	Annual energy yield (GJ/ha)	Annual energy yield (MWh/ha)	Diesel equivalent (litres/day/ha)
Wood *(forestry residues, SRW, thinnings, etc; 30% MC)*	13	2.9 (2 odt)	37	10.3	2.7
Wood *(SRC Willow; 30% MC)*	13	12.9 (9 odt)	167	46	12.2
Miscanthus *(25% MC)*	13	17.3 (13 odt)	225	63	16.5
Wheat straw *(20% MC)*	13.5	3.5 (2.8 odt)	47	13	3.4
Biodiesel *from rapeseed oil*	37	1.1	41	11.3	3
Bioethanol *from sugar beet*	27	4.4	119	33	8.7
Bioethanol *from wheat*	27	2.3	62	17	4.5
Biogas *(60% CH4) from cattle slurry*	30	0.88	26	7.3	1.9
Biogas *(60% CH4) from sugar beet*	30	5.3	159	44	11.6

3. The speech says, for example, 'We must forever conduct our struggle on the high plane of dignity and discipline. We must not allow our creative protest to degenerate into physical violence.' The full version is widely available online, for example at: bitly.com/dream-speech.

4. 'Bigger-than-self problem' is a phrase used by change strategist Tom Crompton to describe a group of problems including climate change, global poverty and biodiversity loss that are in a distinctly different class from those problems where it is 'clearly in an individual's immediate self interest to invest energy and resources

to help tackle'. *Common Cause: A case for working with Cultural Values* (WWF, 2010, wwf.org.uk/change).

5. L. Evans *et al.*, 'Self-interest and pro-environmental behaviour' (*Nature Climate Change*, 2012, bitly.com/evans-paper). Another interesting experiment found that purchasing green-branded products without caring about the underlying issues made people act more selfishly. The researchers propose that this is due to a so-called 'licensing effect', whereby the perception that someone has done a good deed in one circumstance can make us feel less bad about doing a bad deed elsewhere – kind of ethical rebound effect. N. Mazar and C. Zhong, 'Do Green Products Make Us Better People?' (*Psychological Science*, 2010, bitly.com/ethicalrebound).
6. The politician in question is Tony Abbott, who stated recently: 'When I say "there will be no carbon tax under the government I lead", I am telling the truth.' Source: O. Milman, 'Australian politics cools off on climate change – even as the temperature rises' (*The Guardian*, 2013, bitly.com/australia-politics).
7. Naomi Klein's *The Shock Doctrine* (Penguin, 2008) is a fascinating and readable survey of the role of shock in creating rapid change. It opens with an account of free-marketeer Milton Friedman, aged 93, seizing on the aftermath of Hurricane Katrina as the moment to unleash carefully prepared plans to do away with state education. Although Klein's take on the exploitation of chaos is chilling, some of her insights could be put to positive use.

Chapter 15: Capping the carbon

1. Michael Jacobs is one prominent advocate of this idea. See for example 'Climate policy: Deadline 2015' (*Nature*, 2012, bitly.com/mj-deadline).
2. J. Henn, 'Seattle Mayor Orders City to Divest from Fossil Fuels' (350.org, 2012, bitly.com/seattle-fuels).

3. For a summary of the challenges of carbon footprinting products, see M. Berners-Lee, *How Bad are Bananas? The Carbon Footprint of Everything* (Profile Books, 2010).
4. S. Barrett, 'Rethinking Climate Change Governance and its Relationship to the World Trading System' (*The World Economy*, 2011, bitly.com/trade-climate).
5. For an overview of some of the issues see: P. Holmes *et al.*, 'Border Carbon Adjustments and the Potential for Protectionism' (*Climate Policy*, 2011, bitly.com/holmes-2011).
6. J. Aldy *et al.*, 'Climate Change: An Agenda for Global Collective Action'. Prepared for the conference, 'The Timing of Climate Change Policies' (Pew Center on Global Climate Change, 2001, bitly.com/hard-politics).
7. C. Knittel, 'The Energy-Policy Efficiency Gap: Was There Ever Support for Gasoline Taxes?' (NBER Working Paper, 2013, bitly.com/gas-taxes). The study found that 'Polling evidence [from the 1970s] suggests that consumers preferred price controls and rationing and vehicle taxes over higher gasoline taxes.'

Chapter 16: Pushing the right technologies – hard

1. The precise human impact of the Fukushima nuclear incident is subject to ongoing research and debate. One of the most in-depth studies so far estimated that the all-time death toll from radiation may be 130 additional cancer deaths, with a plausible range of 15–1100. (J. Ten Hoeve and M. Jacobson, 'Worldwide Health Effects of the Fukushima Daiichi Nuclear Accident', *Energy & Environmental Science*, 2012.) Nuclear advocates point out that this is a relatively small number compared to the tens of thousands of people killed directly by the unprecedented tsunami that caused the incident – despite the fact that the power plant was old, badly designed, built on a fault line and suffered a catastrophic loss of power. Not everyone accepts these numbers, however, and hundreds of additional casualties are reported to

have died from non-radiation causes during the effort to evacuate the surrounding area.

2. A 'unit' is a kilowatt hour. The Sizewell B figure is for 2011 and taken from EDF (bitly.com/sizewell). The solar figure is based on an average daily output of 12 units a day for a 4kW system and half that for a 2kW system. Source: David MacKay, *Sustainable Energy: Without the Hot Air* (bitly.com/mackay-solar).

3. IEA, *World Energy Outlook 2011*, Box 6.4: The implications of less nuclear power for the 450 Scenario.

4. Telephone interview with Duncan Clark for *The Guardian*. The full quote runs as follows: 'I think if you've seriously looked at ways of making plans that add up you come to the conclusion that you need almost everything and you need it very fast – right now. You need all the credible technologies that can develop at scale. We're not certain which will end up being the lowest cost ones so that's the reason for maintaining a wide portfolio. I don't think anyone serious would say that we only need nuclear – they would say let's have lots of nuclear but we're going to need other technologies too. But similarly I think it's unrealistic to say we could get there solely with renewables … the costs could be very high if you attempted to do that, and there's a lot of very strong resistance to many forms of renewable such as wind farms. So I think it's a sensible thing to do to go for a mixed portfolio.'

5. The 500-year figure is based on current electricity consumption. It was estimated by fast reactor advocate Tom Blees and confirmed by David MacKay, chief scientist at the UK Department for Energy and Climate Change. The reactor in question is the GE Hitachi Prism. For more details see: D. Clark, 'New Generation of Nuclear Reactors Could Consume Radioactive Waste as Fuel' (*The Guardian*, 2012, bitly.com/fast-nuclear).

Chapter 17: Dealing with land and smoke

1. 'Integrated Assessment of Black Carbon and Tropospheric Ozone: Summary for Decision Makers' (UNEP and WMO, 2011, bitly.com/black-carbon).
2. Grantham Research Institute and D. Clark, 'What's Redd and Will It Help Tackle Climate Change?' (*The Guardian*, 2012, bitly.com/redd-faq).
3. Anyone can contribute to the scheme at bitly.com/yasuni-fund. For a recent update on the project see: J. Vidal, 'Can Oil Save the Rainforest?' (*The Observer*, 2013, bitly.com/oil-rainforest). The biodiversity per hectare figure is from: J. Watts, 'World's Conservation Hopes Rest on Ecuador's Revolutionary Yasuni Model' (*The Guardian*, 2012, bitly.com/yasuni-model).
4. See coolearth.org for more information or to donate.
5. In a recent paper exploring this idea, the authors wrote that: 'Both words in the phrase sustainable intensification need to carry equal weight. Intensification, by reducing pressure on land and other resources, underpins sustainability. Equally, food production in the context of a growing population, must ultimately be sustainable if it is to continue to feed people in the future.' T. Garnett and C. Godfray, 'Sustainable Intensification in Agriculture: Navigating a Course through Competing Food System Priorities' (Food Climate Research Network and the Oxford Martin Programme on the Future of Food, 2012, bitly.com/sustainable-intensification).
6. A report by the Manchester University Sustainable Consumption Institute found the public not averse to the concept of factory grown meat substitutes in their 'Lab Chops' diet scenario. See: A. Bows *et al.*, 'What's Cooking? Adaptation and Mitigation in the UK Food System' (SCI, 2012, bitly.com/sci-cooking).

Chapter 18: Making a plan B

1. N. McGlashan *et al.*, 'Grantham Institute for Climate Change Briefing Paper No. 8: Negative Emissions Technologies' (2012, bitly.com/negative-carbon).
2. V. Smetacek *et al.*, 'Deep Carbon Export from a Southern Ocean Iron-Fertilised Diatom Bloom' (*Nature*, 2012, bitly.com/iron-sea).
3. Estimate by Ken Caldeira at the Carnegie Institution of Washington in Stanford, quoted in: M. Marshall, 'Geoengineering with Iron Might Work After All' (*New Scientist*, 2012, bitly.com/iron-might-work).
4. This comment was made by David Keith, a professor of engineering and public policy at Harvard University, in M. Specter, 'The Climate Fixers' (*The New Yorker*, 2012, bitly.com/climate-fixers).

Index

Numbers in italics refer to graphics; 'n' references point to endnotes

INDEX

INDEX

INDEX

INDEX

INDEX

INDEX

INDEX

INDEX

Mike Berners-Lee is a leading expert on carbon emissions, founder of Small World Consulting and author of *How Bad Are Bananas? The Carbon Footprint of Everything*. He is involved in sustainability research across many departments at Lancaster University and has worked on energy and emissions with a wide range of corporate and public sector organisations.

Duncan Clark is a consultant editor on the *Guardian* environment desk, co-founder of digital journalism company Kiln and a visiting researcher at the UCL Energy Institute. He helped set up and run the 10:10 climate campaign, is the author of *The Rough Guide to Green Living* and has edited many books on climate change and related topics.

The David Suzuki Foundation

The David Suzuki Foundation works through science and education to protect the diversity of nature and our quality of life, now and for the future.

With a goal of achieving sustainability within a generation, the Foundation collaborates with scientists, business and industry, academia, government and non-governmental organizations. We seek the best research to provide innovative solutions that will help build a clean, competitive economy that does not threaten the natural services that support all life.

The Foundation is a federally registered independent charity, which is supported with the help of over 50,000 individual donors across Canada and around the world.

We invite you to become a member. For more information on how you can support our work, please contact us:

The David Suzuki Foundation
219–2211 West 4th Avenue
Vancouver, BC
Canada V6K 4S2
www.davidsuzuki.org
contact@davidsuzuki.org
Tel: 604-732-4228
Fax: 604-732-0752

Cheques can be made payable to The David Suzuki Foundation. All donations are tax-deductible.

Canadian charitable registration: (BN) 12775 6716 RR0001
U.S. charitable registration: #94-3204049